BUDDHISM AND ECOLOGY

BUDDHISM And ECOLOGY

Edited by
MARTINE BATCHELOR
KERRY BROWN

MOTILAL BANARSIDASS PUBLISHERS
PRIVATE LIMITED • DELHI

2nd Reprint: Delhi, 2017
First Indian Edition: Delhi, 1994
First Published in the UK by Cassell plc, London

ISBN: 978-81-208-1246-8 (Cloth)
ISBN: 978-81-208-1247-5 (Paper)

MOTILAL BANARSIDASS
41 U.A. Bungalow Road, Jawahar Nagar, Delhi 110 007
8 Mahalaxmi Chamber, 22 Bhulabhai Desai Road, Mumbai 400 026
203 Royapettah High Road, Mylapore, Chennai 600 004
236, 9th Main III Block, Jayanagar, Bengaluru 560 011
8 Camac Street, Kolkata 700 017
Ashok Rajpath, Patna 800 004
Chowk, Varanasi 221 001

Printed in India

by RP Jain at NAB Printing Unit,
A-44, Naraina Industrial Area, Phase I, New Delhi–110028
and published by JP Jain for Motilal Banarsidass Publishers (P) Ltd
41 U.A. Bungalow Road, Jawahar Nagar, Delhi-110007

CONTENTS

SECTION C · MEETING THE GLOBAL CRISIS

ACKNOWLEDGEMENTS

The following illustrations are used by permission:

p. 5: Cultural Printing House of the Tibetan Government.

pp. 26, 32: from Lokesh Chandra, *Buddha in Chinese Woodcuts*, International Academy of Indian Culture.

pp. 44, 113: Serindia Publications, London.

pp. 54, 59: from Myokyo-ni, *Gentling the Bull*, Zen Centre, London.

p. 61: pictures by Giei Satō, from Eshiu Nishimura, *Unsui: A Diary of Zen Monastic Life*, Institute for Zen Studies, University of Hawaii Press.

p. 67: courtesy of the Fogg Art Museum, Harvard University, Cambridge, Massachusetts.

p. 81: from John Landau and Janet Brooke, *Prince Siddhartha*, Wisdom Publications.

p. 92: picture by Mongkol Wongkalasin, from Dr Chatsumarn Kabilsingh, *A Cry from the Forest*, Wildlife Fund Thailand.

pp. 101, 106: Marcelle Hanselaar.

INTRODUCTION

Dharma
Shining
In Leaf Dew[1]

The word *Dharma* means religion. For Buddhists, it is the sacred law, morality and the teachings of the Buddha. It is also all things in nature. Cats, dogs, penguins, trees, humans, mosquitoes, sunlight, leaf dew are all dharmas. So at its very essence, Buddhism can be described as an ecological religion or a religious ecology.

In this book, Buddhist activists, teachers, scholars and leaders from East and West explore what their faith has to offer to humanity as we face a crisis of our own making that threatens all nature.

The book covers three main areas:

1 the environmental perspective found in the Buddhist scriptures;
2 how these teachings have brought and do bring about an ecological lifestyle (and how they have failed to);
3 case studies of contemporary Buddhist responses to the environmental crisis.

The book is divided into three sections accordingly but, true to the Buddhist notion of universal interconnection, each chapter, wherever it is placed, contains elements of all three sections.

The first chapter, selections from the scriptures with short commentaries by Martine Batchelor, outlines Buddhist philosophy. It begins with the principles of love, compassion and respect for all life, which are familiar to the Western mind but which in recent centuries, at least, we have restricted to humans only. Even the law of *karma* (cause and effect) has some place in our thinking although without the universal and inescapable power it is given in Buddhist thought. It is here that we begin to delve into the most challenging aspects of Buddhism for the law of karma ultimately places mind as the first cause. It is the maker and the shaper of our personal destiny and the destiny of the world. In Dr Lily de Silva's detailed exploration of the teachings, this principle is graphically illustrated by the Buddhist legend of a world that physically degenerates as the morality of its inhabitants degenerates.

We are now at the very heart of Buddhist thought and here we discover another concept that we already half-know as we survey our dying world: our birth and existence is dependent on causes outside ourselves, inextricably linking us with the world and denying us any autonomous existence. Indeed, when we think deeply enough, the borders between ourself and the world wash away like water in water. We and all nature are inseparable, entwined, all one. Compassion for others should be as natural and instinctive as compassion for ourselves and our own bodies.

This is perhaps the most striking and difficult idea of Buddhism and the one most misunderstood—that there is no independent, individual self. Yet the individual self is one of the Western world's most cherished beliefs and greatest source of suffering. It is what separates us from the world and causes us to cling to it with the stranglehold of the drowning. In Buddhist thinking, to be enlightened is to awaken from this delusion. As Stephen Batchelor says in his notes towards a Buddhist ecological philosophy: 'The fear that denial of the self would give us no ground to stand on is realized to be in itself groundless, like the discovery we make as children when we find we can swim and are, at that moment, freed from the terror of drowning. Thus the instinctive insistence upon a separate self is seen to provide an utterly false sense of security; for in an undivided world everything miraculously supports everything else.'

These principles lead to the premise of all our contributors that the search for solutions to the global crisis begins within each of us. To transform the world, we must begin by transforming ourselves or, to put it another way, by discovering our true Buddha (enlightened) nature. 'Enlightenment', says Stephen Batchelor, 'is not some mystical state where visions of unearthly bliss unfold but a series of responses to the question: how am I to live in this world?'

In Section B, where we look specifically at Buddhist ways of living in the world, Helena Norberg-Hodge's description of traditional Ladakh is a description of daily life as the path to enlightenment, the 'practice' which the Buddha always emphasized:

> I was beginning to experience the 'wholeness' of this way of life. For the Ladakhis, there are no great distinctions or separations between work and festivity, between human spirituality and attendance to the

> natural environment. . . . Living and working together in a society on this scale, where collective duties and decision-making are usually shared, each member has an overview of the structures and networks of which they are a part. When individuals can experience themselves as part of the community and see the effects of their actions on the whole picture, it is easier both to feel secure and to take responsibility for their own lives.

She also attributes the Ladakhi ability to see things in their 'wholeness' to the fact that 'people spend significant periods of time in a semi-meditative state. Older people in particular recite prayers and mantras as they walk and as they work—even in the middle of conversation.'

In his observations of Japanese ways of life, Wayne Yokoyama also cites meditation as a 'cornerstone' of Buddhist practice and investigates the various forms it takes in Japanese traditions from simply sitting and reciting the Buddha's name to circling a mountain for 1,000 days. Whatever form it takes, meditation is 'a highly practical way of achieving a state of calm in which the seeker perceives reality in perfect clarity'. We return again to the point that we can only act to benefit the world when we have dispelled our own delusions:

> If we are to engage the enemy at its ultimate source, we have to reckon with ourselves and the lives we are leading. For though the hand that fells the tree may not be the same as the gilded one holding the toothpick, they work in league to feed the vanity of the self.

Unlike Japan and Ladakh, Western civilization was becoming industrialized when it met Buddhism. As it was busily fragmenting and separating itself into as many individuals and functions as possible, it was brushed by a religion that denies the reality of this separation. Peter Timmerman looks first at the misunderstandings as Buddhism was absorbed by the nineteenth-century Romantic movement and emerged in Western thought as a religion in which the individual becomes one with nature while somehow becoming more individual in the process.

But as the primacy of the individual and individual desire has continued to grow exponentially in the shadow of the industrializing world, two questions have arisen, says Timmerman: 'How can we deny people their right to self-fulfilment? Yet how can we survive on a planet of ten billion points of infinite greed?' This is the point at which the more challenging aspects of Buddhism present 'a serious alternative basis for environmental thought and action'. Timmerman argues that to be a Buddhist today is a geopolitical act, taking us away from the ethos of the individual and its bondage to the consumer ethic and providing us 'with a working space within which to stand back from our aggressive culture and consider the alternatives. This working space, with its ways of carefully considering and meditating on what we do, is part of what can be called "non-violent thinking".'

It was non-violent thinking that inspired the Sarvodaya movement, the first of our case studies of 'engaged Buddhism' (practical responses to the crisis of our planet based on Buddhist principles). The Sarvodaya movement grew out of a Sri Lankan high school activity in 1958 in which a group of students gave their labour to a destitute village for two weeks and learned how most people in their country lived. Since then, what began as an educational programme has evolved into a self-development programme involving 4,000 towns and villages in Sri Lanka. The inherently political nature of Buddhist action which Peter Timmerman refers to is also apparent to Dr Ariyaratne, the young science teacher who organized the school activity in 1958 and who is now president of the Sarvodaya movement.

> Sarvodaya has committed itself to a dynamic non-violent revolution which is not a transfer of political, economic or social power from one party or class to another, but the transfer of all such power to the people. For that purpose, the individual as well as society must change. . . . In the Buddha's teachings, as much emphasis is given to community awakening and community organizational factors as to the awakening of the individual.

A similar attitude underpins the work of Venerable Ajahn Pongsak, a forest monk in northern Thailand who began working in the 1980s with the villagers of the Mae Soi Valley to reforest and irrigate their rapidly desertifying valley ringed with opium plantations. The Ajahn's work has expanded to nationwide environmental education for other monks and the establishment of an association of 'green' monks who co-operate in their work to protect their local environments.

The Ajahn's activities have prompted police raids and death threats from those in the heroin industry, as well as national and international awards. Despite the warnings to go back to his temple duties, he is in no doubt that his current activities are his duty:

> The true basis of Buddhism is wisdom—the knowledge and understanding of the true worth of nature according to the Natural Law. . . . The Natural Law applies to all life that still must concern itself with the material world, that still must do its duty in and towards society. It is the regular discipline of doing our duty that establishes ethical principles in our daily lives and the general life of society.

Venerable Thich Nhat Hanh, a Vietnamese monk who was nominated for the Nobel Peace Prize by Martin Luther King Jr in 1967, also tasted the violence of police raids when he was working in Singapore to help the boat people as they fled Vietnam. He stresses the importance of 'mindful breathing' and of 'mindfulness verses' that focus and heighten awareness and bring peace and joy to the mind. But he warns:

> Do not avoid contact with suffering or close your eyes before suffering. . . . We should not neglect practices like counting the breath, meditation, and sutra study, but what is the purpose of doing these things? Meditation's purpose is to be aware of what is going on in ourselves and in the world. What is going on in the world can be seen within ourselves and vice versa. Once we see this clearly, we cannot refuse to take a position and act.

One of the great engaged Buddhists of our time, His Holiness the Dalai Lama, is no less adamant about this in his Nobel Peace Prize lecture:

> We must develop a sense of universal responsibility not only in the geographic sense, but also in respect to the different issues that confront our planet. Responsibility does not only lie with the leaders of our countries or with those who have been appointed or elected to do a particular job. It lies with each of us individually. Peace, for example, starts within each one of us. When we have inner peace, we can be at peace with those around us. When our community is in a state of peace, it can share that peace with neighbouring communities, and so on.
>
> . . . It is my dream that the entire Tibetan plateau should become a free refuge where humanity and nature can live in peace and in harmonious balance. . . . Tibet could become a creative centre for the promotion and development of peace.

We have found inspiration from the diverse insights and experiences of our contributors and hope that our readers also learn something of the contribution Buddhism can make, and has made, to ecological thinking and living.

> Master Nansen was asked by a monk: 'Where will the master be gone to in a hundred years' time?' Master Nansen replied: 'I'll be a water-buffalo.' The monk asked: 'May I follow you?' Master Nansen said: 'If you do, bring a mouthful of grass with you!'[2]

Martine Batchelor and Kerry Brown

Notes

1 Issa. From Lucien Stryk and Takashi Ikemoto (ed. and trans.), *Zen Poetry*, Penguin, London, 1981, p. 102.

2 Irmgard Schloegl (trans.), *The Wisdom of the Zen Masters*, Sheldon Press, London, 1975, p. 69.

THE CONTRIBUTORS

Martine Batchelor was born in France and spent ten years as a Zen Buddhist nun in Korea. She translated *The Korean Way of Zen* and has recently completed *The Bottomless Boat: The Zen Journey of Two Women*. She now lives in Britain and leads meditation retreats in England and Europe.

Dr Lily de Silva is a professor in the Department of Buddhist Studies at the Paradenya campus of the University of Sri Lanka.

Stephen Batchelor was a Buddhist monk for ten years studying Buddhist doctrine and the Tibetan language in India, and Zen Buddhism in Korea. He now lives in Britain again and travels extensively through his work as a translator, writer and teacher. He has translated many Tibetan books and is the author of *Alone with Others, The Tibet Guide* and *Faith to Doubt.*

Helena Norberg-Hodge has a degree in linguistics and speaks seven languages. She first went to Ladakh in 1975 where she established ecology and appropriate technology centres. She now lives in Britain again and continues to work fundraising for, and promoting, the Ladakh Project, which she visits regularly. In 1986, she received the Right Livelihood Award.

Wayne Yokoyama is a third-generation Japanese-American who lives in Japan. He is an ordained minister of the Pure Land sect of Buddhism. He is a translator and a language teacher.

Peter Timmerman is a Research Associate of the Institute for Environmental Studies at the University of Toronto in Canada, and the International Federation of Institutes for Advanced Study, also in Toronto. He investigates environmental hazards, including global warming, and sustainable development and is the head of the Secretariat for the Human Dimensions of Global Change Programme. His publications include *Vulnerability, Resilience and the Collapse of Society (A Review of Models for Possible Climatic Applications)* as well as many reports and essays for both government and lay readers.

Dr A. T. Ariyaratne is a former high school science teacher in Sri Lanka. He is the founder and president of the Sarvodaya movement, a self-help village development project which began in 1958 as a school activity. Sarvodaya is now active in 4,000 Sri Lankan towns and villages.

Dr Joanna Macy is an American Buddhist scholar and teacher in movements for social change in both the industrial and the developing world. She spent a year as a participant-observer of the Sarvodaya movement in Sir Lanka and is a board member of Sarvodaya International. As director of Interhelp, she advises organizations campaigning for peace and justice. Her many writings include *Dharma and Development: Religion as Resource in the Sarvodaya Self-Help Movement* and *Despairwork: a Study of the Psychological Impact of the Threat of Nuclear War*.

Phra Ajahn Pongsak Techathammo is the abbot of Wat (temple) Palad in northern Thailand. Since the early 1980s, he has worked with the villagers of the Mae Soi Valley (near Chiang Mai) to help reforest and irrigate their valley. He also runs a Buddhist environmental education programme for monks and laypeople and is the founding head of Thailand's Monks for Preservation and Development of Lives and Environment. In 1990 he received the Global 500 Award.

Kerry Brown is a New Zealander and former journalist. She now lives in Britain and works as religious consultant for the World Wide Fund for Nature. She is an executive-director of the International Sacred Literature Trust and has edited various books on world faiths. She is co-author of the educational scheme *Life Notes: World Music and the Environment*.

Ven. Thich Nhat Hanh is a Vietnamese Buddhist monk. He was the chair of the Vietnamese Buddhist Peace Delegation in Paris during the Vietnam War. Martin Luther King Jr nominated him for the Nobel Peace Prize in 1967. He lives in the south of France where he teaches, writes and gardens. He is the author of *Being Peace*, *The Sun My Heart*, *Present Moment Wonderful Moment*, and many other books.

Tenzin Gyatso, the fourteenth Dalai Lama is the spiritual leader of Tibetans world-wide and the head of the Tibetan government-in-exile. In 1989 he was awarded the Nobel Peace Prize. His books include *My Land and My People*, *Kindness, Clarity and Insight*, *Opening the Eye of New Awareness*, and *Freedom in Exile*.

SECTION A

THE TEACHINGS

1 EVEN THE STONES SMILE

Selections from the scriptures

Edited with commentary by Martine Batchelor

In one of the Buddha's previous lives before he became a Buddha, he lived as a monkey chief high in the Himalayas where clear mountain streams tumbled through ancient forests. The monkey chief's home was a large banyan tree which towered over the other trees, its thick cloud of foliage bowed by a constant abundance of large, juicy, deliciously sweet figs.

One branch of that tree hung over a stream as it passed on its way to the plains and from there to the sea. The far-sighted monkey chief warned his tribe of monkeys, 'unless you prevent that branch from bearing fruit, you will never be able to eat fruit from the other branches'. So every day the monkeys removed any figs that sprouted on the overhanging branch. But eventually a fig began to grow under a large leaf and grew unnoticed until it was a deep, succulent purple and so heavy that it dropped into the dancing waters and swirled away. The stream swept the fruit to the plains where it was scooped out by a woman from the royal harem, as she bathed, and taken to the king. As soon as the king tasted the plump, delicious fig he was overwhelmed by a desire for more. It seemed to him that his very happiness depended on it. He thought: 'If one does not eat such fruits, truly what does one enjoy from one's royalty? But he who has these fruits is really king.'

So the king rallied his troops and they marched forth, following the river across the plain and high into the mountain forest where no humans had ever been before. The forest fell silent before them as they slashed their way through thick twisted branches and flowering vines, frightening even the elephants with the noise of their drums. At last in the distance, like a mass of low-lying clouds, the lord of trees appeared. A sweet mouth-watering smell greeted the army and the king knew this was the tree he sought. Coming closer, he saw hundreds of monkeys in the boughs feasting on the fruits. A wave of anger rose in him; those monkeys were robbing him. 'Drive them away, kill those scoundrels', he bellowed to his army which obediently attacked the monkeys with arrows, clods, sticks and spears.

Far, far up in the highest branches, the monkey chief saw the royal army approach like a billowing wave aroused by a violent wind. He saw the shower of deadly weapons and the pale upturned faces of his terrified tribe as they cried to him. His mind was filled with compassion. He called to his monkeys not to be afraid and then, having resolved to rescue them, climbed to the very top of the tree in line with a mountain peak. Although an ordinary monkey would not have been able to leap even halfway across, his determination was fired by compassion and he sprang to the peak in a single bound. On the slope of the mountain he found a tall, strong deep-rooted cane, the end of which he tied to his foot before jumping back to the tree. But the cane was just short of the distance and the chief only barely managed to grab the nearest branch with his hands. Holding fast, he commanded his tribe to come quickly off the tree. In a state of panic, the monkeys rushed over his body without regard for him and escaped to safety along the cane.

As he was incessantly trodden on by the feet of hundreds of terrified monkeys, the chief's mind held firm but his back weakened, bent and finally broke. Far below, the king and his men watched this extraordinary display of strength, wisdom, self-denial and compassion and were overcome with admiration. The king called to his men to spread out a canopy beneath the mortally injured monkey so they might cut him free of the cane with an arrow shot by their best marksman. After the monkey chief had been safely caught in the canopy and laid on a couch where he was anointed with healing oils, the king approached him. He spoke to the monkey with great respect and curiosity.

'Your honour, what are you to these monkeys or they to you that you show no concern for your own life, but sacrifice your body to save them?'

The dying monkey replied: 'They charged me with the burden of being their ruler and I, bound to them with the affection of a father for his children, accepted this burden.'

The king was even more perplexed: 'The ministers and officials of a king are there to serve the interest of their lord, not the king to serve theirs.'

The monkey replied, 'Yes, that is the law of political wisdom, but to me it seems difficult to follow. It is extremely painful to overlook unbearable suffering even if the sufferer is somebody we do not know. How much more, if those who are suffering are like family to us.'

The monkey's face was bright with joy as he continued, 'True, my body is broken, your Highness, but my mind has come to a state of great soundness, since I alleviated the distress of those whom I ruled for so long and who showed me such affection and reverence. I am happy to bear this pain and the separation from my friends and my approaching death. It seems to me the approach of a high festival.'

The monkey king paused and for a moment the human king thought he had died. But the monkey turned his head to look the man in the eyes and said: 'A king must endeavour to endow with happiness, his army, his animals, his officials and his people, as if he were a father to them. Only then will you enjoy prosperity in this world and the next, illustrious king of men.'

After saying this, the Buddha-to-be left his body and rose to Heaven. The king and his men paid final tribute to the dead leader and then left the forest forever. From that day, the king reigned as the Buddha-to-be had taught him.[1]

LOVE AND COMPASSION

Two of the most important qualities to be developed by Buddhists are loving-kindness and compassion. Loving-kindness is understood as the wish for others to be happy, and compassion as the wish to alleviate suffering. Both start with ourselves, by recognizing the fact of our own suffering and seeking to uproot its causes. Before turning to the plight of others, it is necessary to understand deeply the origins of suffering within ourselves. Such insight can then lead to a genuine capacity to show others the way to freedom from their inner pain. Ultimately loving-kindness and compassion extend to all living things: people, animals, plants, the earth itself.

In this passage from one of the earliest texts of Buddhism,* the Buddha described how a disciple should cultivate loving-kindness:

> [Then let him think:] 'In joy and safety
> Let every creature's heart rejoice,
> Whatever breathing beings there are,
> No matter whether timid or bold,
> With none excepted, long or big
> Or middle-sized or short or thin
> Or thick or those seen or unseen
> Or whether dwelling far or near,
> That are or that yet seek to be.
> Let every creature's heart rejoice.
> Let none betray another's trust
> Or offer any slight at all,
> Or even let them wish in wrath
> Or in revenge each other's ill.'
>
> Thus as a mother with her life
> Will guard her son, her only child,
> Let him extend without bounds
> His heart to every living being.

* It was transmitted orally from the fifth century BCE (Before Common Era) until it was recorded in Sri Lanka in the first century CE (Common Era).

And so with love for all the world
Let him extend without bounds
His heart, above, below, around,
Unchecked, with no ill will or hate.
Whether he stands, or sits, or walks,
Or lies down (while yet not asleep),
Let him such mindfuness* pursue.
This is Holy Abiding here, they say.[2]

From the 8,000-verse Perfection of Wisdom Sutra. Tibetan.

* To be aware of our thoughts and to control them so that they are clear, focused, positive and do not run away with us.

The root of compassion is wisdom. Wisdom is not an introverted 'intellectual' quality but, as the Buddha explains here, gives rise to a spontaneous concern for life:

> In this, *bhikkhu* [monk], a wise person, one of great wisdom, does not intend harm to self, harm to others or harm to both self and others. Thinking in this way, such a person intends benefit for self, benefit for others, benefit for both, benefit for the whole world. Thus is one wise and of great wisdom.[3]

Five hundred years after the Buddha, the Indian Buddhist philosopher Nagarjuna expressed what this insight means for a *bodhisattva*.*

> The essential nature of all bodhisattvas is a great compassionate heart, and all living beings are the object of its compassion.[4]

The eighth-century Buddhist poet Shantideva put it like this:

> In the same way as the hands and so forth
> Are regarded as limbs of the body,
> Likewise, why are living things
> Not regarded as limbs of life?
>
> I should dispel the misery of others
> Because it is suffering, just like my own,
> And I should benefit others
> Because they are living things, just like myself.
>
> When I work in this way for the sake of others
> I should not let conceit or feelings of amazement arise.
> It is just like feeding myself—
> I hope for nothing in return.[5]

When Buddhism travelled across Central Asia into China, the vows of the bodhisattva to help all beings were explained as the appropriate response to the fact that all beings have at one time or another been a parent to us:

> Out of his compassion a child of the Buddha must set living creatures free. Since all male creatures have at one time been our father, they should all be regarded as our father. And since all female creatures have at one time been our mother, they should be regarded as our mother. In each life they have been the ones who have given birth to us.

* Someone who has held back from their final human birth in which they will become a Buddha (i.e. become enlightened and attain eternal release from the world) for the sole purpose of helping others towards enlightenment. The concept of bodhisattvas belongs to the Mahayana tradition (see footnote ***, p. 8).

> Therefore, all living things throughout the six realms* can be considered as our father and mother. So to catch and eat any living creatures is surely equivalent to killing our own parents and eating our old body? Furthermore, the four great elements of earth, water, fire and air are the components of both our own and others' bodies. For these reasons, we should (give life to others) by setting them free.[6]

MORALITY

The morality upon which Buddhist training is based does not come from following the rules without question but out of love and respect for all of life. One of the most simple statements of this principle was given by the Buddha himself:

> Not to commit evil
> But to practise all good
> And to keep the heart pure;
> This is the teaching of the Buddha.[7]

Elsewhere in the scriptures, the Buddha spells out what he means by the right way to lead our lives:

> What is right action? Abstention from killing breathing things, stealing, misconduct in sensual desires: this is called right action.[8]

> There are five trades that a lay follower should not ply. What five? They are: trading in weapons, in breathing things, meat, liquor, and poisons.[9]

One of the clearest explanations of moral conduct is found in the Buddha's advice to his son Rahula, who joined his father's order of monks:

> What do you think about this, Rahula? What is the purpose of a mirror?
>
> Its purpose is reflection, reverend sir.
>
> Even so, Rahula, a physical act should be undertaken only after repeated reflection; a verbal act should be undertaken only after repeated reflection; a mental act should be undertaken only after repeated reflection. If you, Rahula, wish to undertake a physical act, you should reflect on that act so: 'That physical act which I wish to do might lead to the harm of self, and that might lead to the harm of others, and that

* The six realms of Buddhist cosmology are those of the gods, titans, humans, animals, hungry ghosts, and of hell.

> might lead to the harm of both; this physical act is unskilled,* its yield is anguish, its result is anguish'.... But if you, Rahula, while reflecting so, should find: 'That physical act which I wish to do would lead neither to the harm of self nor to the harm of others nor to the harm of both; this act is skilled, its yield is happy, its result is happy'—a physical act like this, Rahula, may be undertaken by you.[10]

Ultimately, the purpose of morality in Buddhism is to create the conditions for the realization of spiritual fulfilment, *nirvana*:

> When a man gives, his merit will increase;
> No enmity can grow in the self-restrained.
> The skilled** shun evil; they attain nirvana
> By ending greed and hatred and delusion.[11]

In Mahayana*** Buddhism, the reflection on morality is extended to consider how everything we use in our daily life is derived from other beings. The Korean Zen Master Ya Un told his monks:

> From the time of ploughing and sowing until the food reaches your mouth and the clothes your body, not only do men and oxen suffer great pains in producing them, but countless insects are also killed and injured. It is improper to benefit in this way from the hardships of others. Even more, how can you endure the thought that others have died in order that you can live? How can you who have an easy life complain of hunger and cold when the farmer is also hungry and cold and the weaving woman is inadequately clothed? A very heavy debt is incurred through wearing fine clothes and eating fine foods.[12]

KARMIC CAUSE AND EFFECT

> For every action we perform we experience a similar result.[13]

The Buddha warned his disciples to remember the traditional Indian doctrine of *karma*. 'Karma' literally means 'actions'. The law of karma states that all our thoughts, words and deeds shape our experi-

* The Buddha stressed the importance of 'skilled' action; that is, action that takes into account the best interests of others given their particular needs, limitations and circumstances.

** Those who are 'skilled' in applying the teachings to the varied circumstances of daily life.

*** Mahayana is the 'Great Vehicle' to carry us across the ocean of suffering to salvation. It is the later of the two main traditions of Buddhism. It stresses loving-kindness and compassion and places greater emphasis on the laity's capacity for enlightenment than does the earlier more monk-centred Theravada tradition. Mahayana, which began in south and north-west India in the first century CE, is now the northern form of Buddhism, found in Tibet, Nepal, China, Korea and Japan.

ences in the future. What each one of us is experiencing now is the outcome of what we have thought, said and done in the past. So, far from being a doctrine of fatalism, the law of karma encourages us to take responsibility for our present situation as well as for how our lives will unfold in the future.

Who does not want to suffer
Should do no evil deeds
Openly or in secret.
Do evil now, then later,
Try though you may to flee it,
Yet surely you will suffer.[14]

A man may plunder as he will.
When others plunder in return,
He, plundered, plunders them again.
The fool believes he is in luck
As long as evil does not ripen;
But when it does, the fool fares ill.
The slayer gets himself a slayer,
The victor finds a conqueror,
The abuser gets himself abused,
The persecutor persecuted;
The wheel of deeds turns round again
And makes the plundered plunderers.[15]

Neither self-made is this puppet,
Nor made by another is this misfortune.
Dependent on a cause it comes to be;
By dissolution of the cause it fades away.
Just as a seed sown in a field will sprout
Due to touch of earth and damp, these two;
Likewise the personality types, elements and senses:
Dependent on a cause, they come to be;
By dissolution of the cause they fade away.[16]

Central to understanding this process of karmic cause and effect is the psychological insight that all action begins in the mind. The quality of life, both in terms of our inner feelings as well as the external situations we find ourselves in, originates from the positive and negative states of mind that motivate our actions. For this reason, disciplined attention to thoughts, feelings and emotion is considered essential to the practice of meditation. The famous opening lines of the Dhammapada express this well:

Mind precedes all things,
Mind is supreme, produced by mind are they.

> If one should speak or act with mind defiled,
> Suffering will follow,
> Just as wheel follows hoof of the drawing ox.
> Mind precedes all things,
> Mind is supreme, produced by mind are they.
> If one should speak or act with purified mind.
> Well-being will follow,
> Like a never-parting shadow.[17]

INTERPENETRATION

The Buddha taught that all things in the universe come into existence, 'arise', as a result of particular conditions. There is no creator God as the first cause, because there is no beginning.

> When that exists, this comes to be; on the arising of that, this arises. When that does not exist, this does not come to be; on the cessation of that, this ceases.[18]

When we understand the nature of ourselves and the world in this way we are freed from the instinctive idea that we and other things somehow exist in our own right, independently and separately from everything else. One of the Buddha's earliest disciples said:

> For one who truly sees the pure and simple arising of phenomena and the pure and simple continuity of conditioned things, there is no fear.
>
> When with wisdom one sees the world as just like grass and wood, not finding any selfishness, one does not grieve with the idea, 'this is not mine'.[19]

The Buddhist philosopher Nagarjuna developed this idea into what is known as the doctrine of 'emptiness'. To assert that all things are empty is not to deny that they exist; it is simply to deny that they are self-existing. For Nagarjuna, the most convincing reason that things are empty of self-existence is that they are dependent upon external conditions to cause them to exist.

> Because there is nothing which is not dependently arisen, there is nothing which is not empty.[20]

When Buddhism came to China, this doctrine of emptiness was further developed to emphasize how things do not merely depend for their existence upon their own immediate set of causes, but upon *everything* in the universe. The Avatamsaka Sutra, which was written in India but became most influential in China, sums up this idea in the following verses:

All lands are my body
And so are the Buddhas living there;
Watch my pores,
And I will show you the Buddha's realm.[21]

Just as the nature of earth is one
While beings each live separately,
And the earth has no thought of oneness or difference
So is the truth of the Buddha.[22]

This insight is beautifully expressed in the Avatamsaka Sutra's metaphor of 'The jewelled net of Indra', explained here by Tu Shun, a sixth-century patriarch of the Hua Yen school:

> Now the celestial jewel net of Kanishka, or Indra, Emperor of gods, is called the net of Indra. This imperial net is made of jewels: because the jewels are clear, they reflect each other's images, appearing in each other's reflections upon reflections, *ad infinitum*, all appearing at once in one jewel and in each one it is so—ultimately there is no going or coming. . . .
>
> If you sit in one jewel, then you are sitting in all jewels in every direction, multiplied over and over. Why? Because in one jewel there are all the jewels. If there is one jewel in all jewels, then you are sitting in all jewels too. And the reverse applies to the totality if you follow the same reasoning. Since in one jewel you go into all the jewels without leaving this one jewel, so in all jewels you enter one jewel without leaving this one jewel.[23]

In this way the Chinese Buddhists realized how all things are intimately connected with each other. They called this the doctrine of the 'mutual interpenetration and interfusion of all phenomena'. Uisang, a monk from Korea, put it in these lines of verse:

Since Dharma-nature is round and interpenetrating, it is without any sign of duality.
All dharmas (phenomena) are unmoving and originally calm.
No name, no form; all (distinctions) are abolished.
It is known through the wisdom of enlightenment, not by any other level.
The true-nature is extremely profound, exceedingly subtle and sublime.
It does not attach to self-nature, but takes form following (causal) conditions.
In one is all, in many is one.
One is identical to all, many is identical to one.
In one particle of dust are contained the ten directions.*
And so it is with all particles of dust.[24]

* North, south, east, west; north-east, south-east, south-west, north-west; zenith, nadir (up and down).

This philosophy of interpenetration had a great influence on the Zen tradition, where it found a less abstract expression. Like many others, the Japanese Zen master Dogen turned his enlightened attention to the ordinary details of everyday life:

> There are myriads of forms and hundreds of grasses throughout the entire earth, and yet each grass and each form itself is the entire earth.[25]

> It is not only that there is water in the world, but there is a world in water. It is not just water. There is also a world of living things in clouds. There is a world of living things in the air. There is a world of living things in fire. There is a world of living things on earth. There is a world of living things in the phenomenal world. There is a world of living things in a blade of grass. There is a world of living things in one staff. Whenever there is a world of living things, there is a world of Buddha ancestors. You should examine the meaning of this.[26]

This way of experiencing the world in which we live is uncannily similar to much of the language of contemporary ecological thinking. The contemporary Vietnamese monk and poet Thich Nhat Hanh encourages his students to break out of their self-centredness and understand the interconnection of all living things.

> When we look at a chair, we see the wood, but we fail to observe the tree, the forest, the carpenter, or our own mind. When we meditate on it, we can see the entire universe in all its interwoven and interdependent relations in the chair. The presence of the wood reveals the presence of the tree. The presence of the leaf reveals the presence of the sun. The presence of the apple blossoms reveals the presence of the apple. Meditators can see the one in many, and the many in the one. Even before they see the chair, they can see its presence in the heart of living reality. The chair is not separate. It exists only in its interdependent relations with everything else in the universe. It *is* because all other things *are*. If it is *not*, then all others are *not* either.[27]

NATURE

Concern for the welfare of the natural world has been an important element throughout the history of Buddhism. Recognition that human beings are essentially dependent upon and interconnected with their environments has given rise to an instinctive respect for nature. Although Buddhists believe humans have a unique opportunity to realize enlightenment, which other creatures do not, they have never believed humanity is superior to the rest of the natural world. This respect for nature is clearly revealed in the following exchange between the Buddha and his disciple Maha Moggallana when the monks' custom of receiving their daily food as charity from local people was undermined by famine.

> The venerable Maha Moggallana went to the Blessed One. He said: 'Lord, alms food is hard to get in Veranja now. There is a famine and food tickets have been issued. It is not easy to survive even by strenuous gleaning. Lord, this earth's under-surface is rich and as sweet as pure honey. It would be good if I turned the earth over. Then the bhikkhus (monks) will be able to eat the humus that water plants live on.'
>
> 'But, Moggallana, what will become of the creatures that depend on the earth's surface?'
>
> 'Lord, I shall make one hand as broad as the Great Earth and get the creatures that depend on the earth's surface to go on to it. I shall turn the earth over with the other hand.'
>
> 'Enough, Moggallana, do not suggest turning the earth over. Creatures will be confounded.'[28]

In another exchange the Buddha compares the kind of religious ceremony he approves of with those common at his time:

> In this sacrifice, Brahmin, no bulls were slain, no goats or sheep, no cocks and pigs, nor were various living beings subjected to slaughter, nor were trees cut down for sacrificial posts, nor were grasses mown for the sacrificial grass, and those who are called slaves or servants or workmen did not perform their tasks for fear of blows or threats, weeping and in tears. But those who wanted to do something did it, those who did not wish to did not: they did what they wanted to do, and not what they did not want to do. The sacrifice was carried out with ghee, butter, curds, honey, molasses.[29]

The natural environment, uninhabited by humanity, was also respected as the ideal place for cultivating spiritual insights. Even in eighth-century India, dwelling in nature was clearly preferable for Shantideva to life in a monastery or town:

> When shall I come to dwell in forests
> Amongst the deer, the birds and the trees,
> That say nothing unpleasant
> And are delightful to associate with?[30]

Milarepa, Tibet's great yogi-saint, was also fond of praising the benefits of living alone in the wild:

> This is a delightful place, a place of hills and forests.
> In the mountain-meadows, flowers bloom;
> In the woods dance the swaying trees!
> For monkeys it is a playground.
> Birds sing tunefully,

Bees fly and buzz,
And from day until night the rainbows come and go.
In summer and winter falls the sweet rain,
And mist and fog roll up in fall and spring.
At such a pleasant place, in solitude,
I, Milarepa, happily abide,
Meditating upon the void-illuminating Mind.[31]

And Zen master Dogen too:

Even *zazen* hours* advance. Sleep hasn't come yet.
More and more I realize mountain forests are good for efforts in the way
Sounds of the valley brook enter the ears, moonlight pierces the eyes.
Outside this, not one further instant of thought.[32]

In such settings it is not surprising to find natural phenomena used as metaphors to make spiritual insights comprehensible:

> A fish swims in the ocean, and no matter how far it swims there is no end to the water. A bird flies in the sky, and no matter how far it flies there is no end to the air. However, the fish and the bird have never left their elements. When their activity is large their field is large. When their need is small their field is small. Thus, each of them totally covers its full range, and each of them totally experiences its realm. If the bird leaves the air it will die at once. If the fish leaves the water it will die at once.
>
> Know that water is life and air is life. The bird is life and the fish is life. Life must be the bird and life must be the fish.[33]

Especially in East Asia, Buddhist meditators would often prefer to express themselves in poems rather than philosophical utterances:

Cherry blossoms?
In these parts
grass also blooms.[34]

Buddha law
Shining
In leaf dew.[35]

Fireflies
Entering my house
Don't despise it.[36]

Clouds of mosquitoes—
It would be bare
Without them.[37]

* Hours for seated meditation.

Winding back and forth
Among green trees
The golden shuttle
Of the oriole
Weaves silk
The colour of spring.
A monk sits
Dozing . . .
Even the stones smile.[38]

CONCLUDING REMARKS

In their attempts to understand the meaning of human life, Buddhists throughout the ages have discovered that 'the truth that sets the heart free' is not found in some metaphysical reality outside the place and time in which each of us lives and breathes. The Buddha and his followers have always been wary of any attempt to endow human existence with a higher or divine purpose—the workings of a remote all-powerful God or eternal soul or even the modern concept of 'progress'. The spiritual meaning of life is to be found right here in the midst of this network of relationships we call 'life'. A modern American Buddhist abbot, Ajahn Sumedho, explains:

> Once you see what it is all about, you really want to be very careful about what you do and say. You can have no intention to live life at the expense of any other creature. One does not feel that one's life is so much more important than anyone else's. One begins to feel the freedom and the lightness in that harmony with nature rather than the heaviness of exploitation of nature for personal gain. When you open the mind to the truth, then you realize there is nothing to fear. What arises passes away, what is born dies, and is not self—so that our sense of being caught in an identity with this human body fades out. We don't see ourselves as some isolated, alienated entity lost in a mysterious and frightening universe. We don't feel overwhelmed by it, trying to find a little piece of it that we can grasp and feel safe with, because we feel at peace with it. Then we have merged with the Truth.[39]

Finally, the sense of urgency that our present social, political and ecological crises provoke is summed up in this statement by the Buddhist monk Tenzin Gyatso, fourteenth Dalai Lama of Tibet and winner of the 1989 Nobel Peace Prize:

> Today more than ever before, life must be characterized by a sense of Universal Responsibility, not only nation to nation and human to human, but also human to other forms of life.[40]

Notes

All translations are published by the Pali Text Society, London, unless another publisher is named.

Abbreviations used are as follows:

A	Anguttara Nikaya
D	Digha Nikaya
Dh	Dhammapada
J	Jataka
M	Majjhima Nikaya
S	Samyutta Nikaya
Sn	Sutta-nipata
Thag	Theragatha
U	Udana
Vin	Vinaya Pitaka

1 From the Jatakamala (stories of the Buddha's previous lives), adapted by Kerry Brown.
2 Sn I, 8.
3 A II, 179.
4 Nagarjuna, *Elegant Sayings*, ed. Sakya Pandit, Dharma Publishing, Emeryville, 1977.
5 Shantideva, *Bodhicaryavatara* VIII, 114. Trans. Stephen Batchelor, *A Guide to the Bodhisattva's Way of Life*, Library of Tibetan Works and Archives, Dharamsala, 1979, p. 118.
6 The Bodhisattva Precepts no. 20. Unpublished translation by Martine Batchelor.
7 The Dhammapada.
8 S XLV, 8; D 22.
9 A V, 177.
10 M II, pp. 89–90.
11 D 16; U VIII.
12 Ya Un Sunim, 'On Self-Admonition'. Unpublished translation by Martine Batchelor.
13 Gunaprabha, Vinaya Sutra.
14 U V, 4.
15 S III, 14–15.
16 S I, p. 134.
17 Dh 1–2.
18 M I, p. 134.
19 Thag 716–17.
20 Nagarjuna, *Madhyamikakarika* XXIV, 19.
21 Avatamsaka Sutra. Trans. Thomas Cleary, *The Flower Ornament Scripture*, Vol. 1, Shambhala, Boston, 1984, p. 184.
22 Avatamsaka Sutra. Cleary. *Ibid*. p. 302.
23 Tu Shu, 'Cessation and Contemplation in the Five Teachings of the Huayen'. Thomas Cleary (ed.), *Entry into the Inconceivable*, University of Hawaii Press, Honolulu, 1983, p. 66.
24 Uisang, 'Ocean Seal of Huayen Buddhism'.
25 Dogen. *Moon in a Dewdrop*, ed. Kazuaki Tanahashi, Element Books, Shaftesbury, 1988, p. 15.

26 Dogen. *Ibid.*, pp. 106–7.
27 Thich Nhat Hanh, *The Sun My Heart*, Parallax Press, Berkeley, 1988, p. 90.
28 Vin, Sutta-vighanga, para. I.
29 Kutadanta Sutta 5, 18. Trans. Maurice Walshe, *Thus I Have Heard*, Wisdom Publications, Boston, 1988, p. 138.
30 Shantideva. *Bodhicaryavatara* VIII, 25. Translation (see note 5), p. 103.
31 Milarepa. Trans. Garma C. C. Chang, *The Hundred Thousand Songs of Milarepa*, Vol. 1, Shambala, Boston, 1989, pp. 74–5.
32 Dogen. *Moon in a Dewdrop*, p. 216.
33 Dogen. *Ibid.*, pp. 71–2.
34 Issa. From Lucien Stryk and Takashi Ikemoto (ed. and trans.), *Zen Poetry*, Penguin, London, 1981, p. 100.
35 Issa. *Ibid.*, p. 102.
36 Issa. *Ibid.*, p. 106.
37 Issa. *Ibid.*, p. 107.
38 Kusan Sunim, *The Way of Korean Zen*, John Weatherhill, Tokyo/New York, 1985, p. 59.
39 Ajahn Sumedho. Cited in Santacitto Bhikkhu (ed.), *Buddha-Nature*, WWF, London, 1989.
40 HH the Dalai Lama. Cited in Martin Palmer, Anne Nash and Ivan Hattingh (eds), *Faith and Nature*, Century, London, 1988.

2 THE HILLS WHEREIN MY SOUL DELIGHTS

Exploring the stories and teachings

Lily de Silva

In our search for pleasure and affluence modern humanity has exploited nature without any moral restraint to such an extent that nature has been rendered almost incapable of sustaining healthy life. Invaluable gifts of nature, such as air and water, have been polluted with disastrous consequences. Humanity is now searching for ways and means of overcoming the pollution problem as our health is also alarmingly threatened. If we are to act with a sense of responsibility to the natural world, to our fellow human beings and to future generations, we must find an appropriate environmental ethic. Hence our search for wisdom and new attitudes in that neglected area of knowledge, religion.

As Buddhism is a fully-fledged philosophy of life reflecting all aspects of experience, it is possible to find enough material in the Pali Canon* to delineate the Buddhist attitude towards nature.

NATURE AS DYNAMIC

According to Buddhism, changeability, expressed by the Pali term *anicca*, is one of the basic principles of nature. The world is dynamic and is therefore defined as that which disintegrates (*lujjati ti loko*). There are no static and stable 'things'; there are only ever-changing, ever-moving processes. Solidity, liquidity, heat and mobility, recognized as the building blocks of nature, are all ever-changing phenomena. Rain is a good example to illustrate this point. Though we use a noun called 'rain' which appears to name a *thing*, rain is nothing but the *process* of drops of water falling from the skies. Apart from this

* The Buddhist scriptures as they were first recorded in the first century CE in Sri Lanka. They were written down in the language now known as Pali (lit. 'scripture') which was a dialect of the Magadha kingdom of India.

process, the activity of raining, there is no rain. Even the most solid looking mountains and the very earth that supports everything on it are not beyond this inexorable law of change. One *sutta*★ explains how the massive king of mountains, Mount Sineru, which is rooted in the great ocean to a depth of 84,000 leagues and which rises above sea level to another great height of 84,000 leagues and which is the classical symbol of stability and steadfastness—also is destroyed by heat, without leaving even ashes, at a time when multiple suns appear in the sky.[1] Thus change is the very essence of nature.

MORALITY AND NATURE

The world passes through alternating cycles of evolution and dissolution, each of which lasts for a very long time. Though change is inherent in nature, Buddhism believes that natural processes are affected by the morals of humanity.

This is vividly illustrated in the Agganna Sutta,[2] which tells the Buddhist legend about the evolution of the world. In the legend, the first beings were self-luminous, subsisted on joy and flew through the skies, until greed entered their minds. This caused the gradual loss of their radiance and of their ability to subsist on joy and to fly. The moral decline affected the external environment as well. At that time, the entire earth was covered by a very flavoursome, fragrant substance similar to butter. When the first beings began eating this substance with increasing greed, their subtle bodies became coarser and coarser while the flavoursome substance itself started to diminish. As the bodies of the beings solidified, differences of form appeared; some were beautiful while others were homely. Thereupon conceit developed in the beautiful ones and they started looking down upon the others. As a result of these moral blemishes, the delicious edible earth-substance completely disappeared. In its place there appeared edible mushrooms and later another kind of edible creeper. In the successive generations who subsisted on this vegetation, sexual differentiation developed and the former method of spontaneous birth was replaced by sexual reproduction.

Self-growing rice appeared on earth and, out of laziness, people began to hoard food rather than collect each meal. As a result, the growth rate of food could not keep pace with demand. Therefore

★ A single teaching by the Buddha given in response to a specific need or theme of daily life. The sutta (also known as 'sutra') are collected in the Sutta-Pitaka, one of the three pitakas (baskets) that make up the Pali canon or Tripitaka ('three baskets').

land had to be divided among families. After private ownership of land became the order of the day, the more greedy people started robbing from others' land. When they were detected, they denied that they had stolen. Thus, through greed, vices such as stealing and lying developed in society. To curb the wrong-doers and punish them, a king was elected by the people and so the original simple society became much more complex. It is said that this moral degeneration of humanity had adverse effects on nature. The richness of the earth diminished and self-growing rice disappeared. People had to till the land and cultivate rice. This rice grain was coated in chaff; it needed cleaning before it could be eaten.

The point I wish to emphasize by citing this legend is that Buddhism believes that, although change is inherent in nature, humanity's moral deterioration accelerates and shapes the changes bringing about circumstances which are adverse to human well-being and happiness.

The Cakkavattisihanada Sutta predicts the future course of events when human morals degenerate further.[3] Gradually people's health will deteriorate so much that life expectancy will decrease until the average human life-span is reduced to ten years and marriageable age to five years. At that time all delicacies such as ghee, butter, honey, etc. will have disappeared from the earth; what is considered the poorest, coarsest food today will become a delicacy.

According to a discourse of the Buddha in the Anguttara Nikaya, when lust, greed and wrong values grip the heart of humanity and immorality becomes widespread in society, timely rain does not fall. When timely rain does not fall, crops fall victim to pests and plant diseases. Through lack of nourishing food the human mortality rate rises.[4]

Thus several suttas from the Pali Canon show that early Buddhism believes there is a close relationship between human morality and the natural environment. This idea has been systematized in the theory of the five natural laws (*panca niyamadhamma*) in the later commentaries.[5] According to this theory, in the cosmos there are five natural laws of forces at work, namely *utuniyama* (lit. 'season-law'), *bijaniyama* (lit. 'seed-law'), *cittaniyama* (lit. 'mind-law), *kammaniyama* (lit. 'action-law) and *dhammaniyama* (lit. 'phenomenal/universal-law'). They can be translated as physical laws, biological laws, psychological laws, moral laws and causal laws, respectively. While the first four laws operate within their respective spheres, the law of causality operates within each of them as well as between them.

This means that the physical elements, i.e. earth, water, air conditions of any given area affect the growth and development of its biological component, i.e. flora and fauna. This in turn influences the thought patterns of the people who interact with the flora and fauna. These thought patterns determine moral standards. The opposite process of interaction is also possible. The morals of humanity influence not only the psychological make-up of the people but the biological and physical environment of the area as well. Thus the five laws demonstrate that humanity and nature are bound together in a reciprocal relationship with changes in one necessarily causing changes in the other.

The commentary on the Cakkavattisihanada Sutta goes on to explain the pattern of mutual interaction in more detail.[6] When humanity is demoralized through greed, famine is the natural outcome; when moral degeneration is due to ignorance, epidemic is the inevitable result; when hatred is the demoralizing force, widespread violence is the ultimate outcome. If and when humanity realizes that large-scale devastation has taken place as a result of its moral decline, a change of heart takes place among the few surviving human beings. As morality is renewed, conditions improve through a long period of cause and effect and humanity again starts to enjoy gradually increasing prosperity and longer life. The world, including nature and humanity, stands or falls with the type of moral force at work. If immorality grips society, people and nature deteriorate; if morality reigns, the quality of human life and nature improves. Thus greed, hatred and delusion produce pollution within and without. Generosity, compassion and wisdom produce purity within and without. This is one reason the Buddha has pronounced that the world is led by the mind, *cittena niyati loko.*[7]

HUMAN USE OF NATURAL RESOURCES

For survival, humans depend on nature for their food, clothing, shelter, medicine and other needs. For maximum benefit, humans have to understand nature so that they can use natural resources while living harmoniously with nature. By understanding the working of nature—for example, the seasonal rainfall pattern, methods of conserving water by irrigation, the soil types, the physical conditions required for growth of various food crops—humans can get better returns from their farming. But this learning has to be accompanied by moral restraint if we are to enjoy the benefits of natural resources for a long time. Humanity must learn to satisfy its needs and not feed its greeds. The resources of the world are not unlimited whereas human greed knows neither limit nor satiation.

These days ostentatious consumerism is accepted as the order of the day. One writer says that within forty years Americans alone have consumed the same quantity of natural resources as had been consumed by all humanity in the previous 4,000 years.[8] The vast non-replenishable resources of fossil fuels which took millions of years to form have been consumed within a couple of centuries to the point of near exhaustion. This consumerism has given rise to an energy crisis on the one hand and a pollution problem on the other. Modern humanity in unbridled greed for pleasure and acquisition of wealth is killing the goose that laid the golden egg.

Buddhism tirelessly advocates the virtues of non-greed, non-hatred and non-delusion in all human pursuits. Greed breeds sorrow and unhealthy consequences. Contentment is a highly-praised virtue in Buddhism.[9] The person leading a simple life with few wants easily satisfied is held up as an exemplary character.[10] Miserliness[11] and wastefulness[12] are equally deplored as two degenerate extremes. Wealth should only be a means to an end; it is for the satisfaction of human needs. Hoarding is a senseless anti-social habit comparable to the attitude of the dog in the manger. The vast hoarding of wealth in some countries and the methodical destruction of large quantities of agricultural produce to keep the market prices from falling, while half the world is dying of starvation, is a tragic paradox of the present affluent age.

Buddhism commends frugality as a virtue in its own right. The disciple Ananda explained to King Udena the thrifty uses of robes by the monks. When new robes were received the old robes were used as coverlets, the old coverlets as mattress covers, the old mattress covers as rugs, the old rugs as dusters, and the old tattered dusters are kneaded with clay and used to repair cracked floors and walls.[13] Thus nothing was wasted. Those who waste are derided as 'wood-apple eaters'.[14] A man shakes the branch of a wood-apple tree and all the fruits, ripe as well as unripe, fall. The man collects only what he wants and leaves the rest to rot. Such a wasteful attitude is certainly deplored in Buddhism as not only anti-social but criminal. The excessive exploitation of nature as carried out today would certainly be condemned by the Buddha in the strongest possible terms.

Buddhism advocates a gentle non-aggressive attitude towards nature. According to the Sigalovada Sutta a householder should accumulate wealth as a bee collects nectar from a flower.[15] The bee harms neither the fragrance nor the beauty of the flower, but gathers nectar to turn it into sweet honey. Similarly, a human being is expected to make legitimate use of nature so that s/he can rise above nature and realize his or her innate spiritual potential.

ATTITUDE TOWARDS ANIMAL AND PLANT LIFE

The well-known Five Precepts (*panca sila*) form the minimum code of ethics that every lay Buddhist is expected to follow. The first precept is non-injury to life. It is explained as the casting aside of all forms of weapons and being careful not to deprive a living thing of life. The Buddhist layperson is also expected to abstain from trading in meat.[16]

Buddhist monks and nuns have to follow an even stricter code of ethics than the layperson. They must abstain from practices which would even unintentionally harm living creatures. For instance, the Buddha made a rule against travelling during the rainy season because of possible injury to worms and insects that come to the surface in wet weather.[17] The same concern for non-violence prevents a monk from digging the ground.[18] Once a monk, who was a potter before he was ordained, built himself a clay hut and set it on fire to give it a fine finish. The Buddha strongly objected to this as so many living creatures would have been burnt in the process. The hut was taken down on the Buddha's instructions to prevent it from setting a bad example for later generations.[19] The scrupulous non-violent attitude towards even the smallest living creatures prevents the monks and nuns from drinking unstrained water.[20] It is no doubt a sound hygienic habit, but what is noteworthy is the reason which prompts the practice, namely, sympathy for other creatures.

In its positive sense the first precept of non-injury means the cultivation of compassion and sympathy for all living beings.[21] The Karaniyametta Sutta prescribes the practice of *metta*, 'loving-kindness' towards all creatures, timid and bold, long and short, big and small, minute and great, visible and invisible, near and far, born and awaiting birth.[22] Just as our own life is precious to us, so is the life of another precious to it. Therefore reverence must be cultivated towards all forms of life.

The Nandivisala Jataka illustrates how kindness should be shown to domesticated animals.[23] Even a wild animal can be tamed with kind words. Parileyya was a wild elephant who waited on the Buddha when he spent time in the forest away from the monks.[24] Another elephant, the infuriated Nalagiri, was tamed by the Buddha with no other miraculous power than the power of loving-kindness.[25] Human and beast can live and let live without fear of one another if only humans cultivate sympathy and regard all life with compassion.

The understanding of kamma (or karma) and rebirth also prepares the Buddhist to adopt a sympathetic attitude towards animals.

According to this belief humans can be reborn as animals. So it is possible that our dead relatives are now living as animals. Therefore, it is only right that we should treat animals with kindness and sympathy. The Buddhist concept of merit also encourages a gentle non-violent attitude towards living creatures. It is said that if someone throws dish-washing water into a pool where there are small creatures so that they can feed on the tiny particles of food washed away, that person accumulates spiritual merit even by such trivial generosity.[26] According to the Macchuddana Jataka, in a previous life, the Buddha-to-be threw his leftover food into a river to feed the fish, and by the power of that merit he was saved from an impending disaster.[27] Thus kindness to animals, be they big or small, is a source of merit—merit that human beings need to improve their lot in the cycle of rebirths and to approach the final goal of *Nibbana* (also known as 'Nirvana').

Buddhism also expresses a gentle non-violent attitude towards the vegetable kingdom which provides us with all necessities of life. It is said that we should not even break the branch of a tree that has given us shelter.[28]

Among Buddhists, large, old trees are particularly revered. The attitude, which is a legacy of pre-Buddhist animism, does not violate the belief system of Buddhism. The trees are called *vanaspati* in Pali, meaning 'lords of the forests'.[29] The deference to trees is further strengthened by the fact that huge trees such as the ironwood, the sala and the fig tree are acknowledged as Bodhi trees, trees under which former Buddhas attained enlightenment.[30] It is well known that the fig species *ficus religiosa* is held as an object of great veneration in the Buddhist world as the tree under which the Buddha attained enlightenment.

The construction of parks and pleasure groves for public use is considered a great deed that gains much spiritual merit.[31] Sakka, the lord of gods, is said to have reached this position as a result of service such as the construction of parks, pleasure groves, ponds, wells and roads.[32]

The open air, natural habitats and forest trees have a special fascination for the Eastern mind as symbols of spiritual freedom. Home life is regarded as a fetter that keeps a person in bondage and misery. Renunciation is like the open air, nature unhampered by human activity.[33] The chief events in the life of the Buddha took place in the open air. He was born in a park at the foot of a tree in Kapilavatthu; he attained enlightenment in the open air at the foot of the Bodhi tree in Bodhgaya; he began his missionary activity in the open air in the sala grove of the Mallas in Pava. The Buddha's constant advice to his disciples was to resort to natural habitats such as the forests. There, undisturbed by human activity, they could devote themselves to meditation.[34]

ATTITUDE TOWARDS POLLUTION

Environmental pollution has assumed such vast proportions today that humanity has been forced to recognize the presence of an ecological crisis. We can no longer turn a blind eye to the situation as we are already threatened with new pollution-related diseases. Pollution to this extent was unheard of during the time of the Buddha. But there is sufficient evidence in the scriptures to provide insight into the Buddhist attitude towards pollution. Cleanliness, both in the person and in the environment, was highly commended. Several rules prohibit monks from polluting green grass and water with saliva, urine and faeces.[35] These were the common agents of pollution known during the Buddha's day. Rules about keeping the grass clean were prompted by ethical and aesthetic considerations as well as the fact that it is food for many animals. Water, whether in a river, pond or well, was for public use and each individual had to use it with proper care so that others who followed could use it with the same degree of cleanliness.

Today, noise is recognized as a serious personal and environmental pollutant troubling everyone to some extent. It causes deafness, stress and irritation, breeds resentment, saps energy and lowers efficiency.[36] The Buddha did not hesitate to voice his stern disapproval of noise whenever the occasion arose.[37] Once he ordered a group of monks to leave the monastery for noisy behaviour.[38] Even in their choice of monasteries the presence of undisturbed silence was an important quality the Buddha and his disciples looked for.[39] Silence invigorates those who are pure at heart and raises their efficiency for meditation. But silence overawes those who are impure with ignoble impulses of greed, hatred and delusion. The Bhayabherava Sutta beautifully illustrates how even the rustle of a falling twig in the quiet of the forest sends tremors through an impure heart.[40] This may perhaps account for the present craze for constant auditory stimulation with transistors and cassettes. The moral impurity caused by greed, avarice, acquisitive instincts and aggression has made people fear silence which lays bare the reality of self-awareness. They prefer to drown themselves in loud music.

The psychological training of the monks is so advanced that they are expected to cultivate a taste not only for external silence, but for inner silence of speech, desire and thought as well. The sub-vocal speech, the inner chatter that goes on constantly within us in our waking life, is expected to be silenced through meditation.[41] The sage who succeeds in completely quelling this inner speech is called a *muni*, a silent one.[42] The inner silence is maintained even when speaking!

It is worth noting as well the Buddhist attitude to speech. Moderation in speech is considered a virtue, as one can avoid four unwholesome vocal activities, namely, falsehood, slander, harsh speech and frivolous talk. In its positive aspect, moderation in speech paves the way to self-awareness. Buddhism commends speaking at the appro-

Fourteenth-century Chinese woodblock.

priate time, speaking the truth, speaking gently, speaking what is useful, and speaking out of loving-kindness; the opposite modes of speech are condemned.[43] The Buddha's general advice to the monks regarding speech was to discuss the Dhamma (the teachings, the universal law) or maintain noble silence.[44]

NATURE AS BEAUTIFUL

The Buddha and his disciples regarded natural beauty as a source of great joy and aesthetic satisfaction. The saints who purged themselves of sensuous worldly pleasures responded to natural beauty with a detached appreciation. Many poets derive inspiration from nature because of the sentiments it arouses in their hearts; they become emotionally involved with nature. For instance, they may compare the sun's rays passing over the mountain tops to the blush on a sensitive face; they may see a tear in a dew drop; the lips of their beloved in a rose petal. But the appreciation of the saint is quite different. The saint appreciates nature's beauty for its own sake and derives joy unsullied by sensuous associations and self-projected ideas. The simple spontaneous appreciation of nature's exquisite beauty is expressed by the Elder Mahakassapa in the following words:

> Those upland glades delightful to the soul,
> Where the Kaveri spreads its wildering wreaths,
> Where sound the trumpet-calls of elephants:
> Those are the hills where my soul delights.
>
> Those rocky heights with hue of dark blue clouds
> Where lies embossed many a shining lake
> Of crystal-clear, cool waters, and whose slopes
> The 'herds of Indra'* cover and bedeck:
> Those are the hills wherein my soul delights.
>
> Fair uplands rain-refreshed, and resonant
> With crested creatures' cries antiphonal,
> Lone heights where silent sages oft resort:
> Those are the hills wherein my soul delights.[45]

CONCLUSION

In the modern age people have become alienated from themselves

* Indra is the Indian sky-god whose herds of elephants are seen as clouds.

and nature. When science started unveiling the secrets of nature one by one, humanity gradually lost faith in theistic religions.* Consequently moral and spiritual values were also discarded. Since the Industrial Revolution and the consequent acquisition of wealth through technological exploitation of nature, humanity has become more and more materialistic. The pursuit of sensory pleasures and the acquisition of possessions have become ends in themselves. The senses dominate people and they are slaves to their insatiable passions. (Incidentally the sense faculties are known in Pali as *indriyas* or lords because they control a person unless s/he is sufficiently vigilant to keep control of them.) Thus men and women have become alienated from themselves as they abandon themselves to sensual pleasures and acquisitive instincts.

In our greed for more and more possessions, we have adopted a violent and aggressive attitude towards nature. Forgetting that we are a part and parcel of nature, we exploit it with unrestrained greed, thereby alienating ourselves from it as well. The result is the deterioration of humanity's physical and mental health on the one hand, and the rapid depletion of non-replenishable natural resources and environmental pollution on the other. These results remind us of the Buddhist teachings in the suttas discussed above, which maintain that the moral degeneration of humanity leads to a decrease in life-span and the depletion of natural resources.

Moral degeneration is a double-edged weapon, it has adverse effects on humanity's mental and physical well-being as well as on nature. Depletion of vast resources of fossil fuels and forests has given rise to a very severe energy crisis. It cannot be emphasized too strongly that such rapid depletion of non-renewable natural resources within less than two centuries, an infinitesimal fraction of the millions of years taken for them to form, is due to modern society's inordinate greed and acquisitiveness. A number of simple ancient societies had advanced technological skills, as is apparent from their vast sophisticated irrigation schemes designed to meet the needs of large populations. Yet they survived in some countries for over 2,000 years without such problems as environmental pollution and depletion of natural resources. This was no doubt due to the philosophy which inspired and formed the basis of these civilizations.

In the present ecocrisis humanity has to look for radical solutions. 'Pollution cannot be dealt with in the long term on a remedial or cosmetic basis or by tackling symptoms: all measures should deal

* Religions based on belief in a Creator God who is involved in the workings of the world but is also above and beyond it.

with basic causes. These are determined largely by our values, priorities and choices.'[46] The human race must reappraise its value system. The materialism that has guided our lifestyle has landed us in very severe problems. Buddhism teaches that mind is the forerunner of all things, mind is supreme. If we act with an impure mind, i.e. a mind sullied with greed, hatred and delusion, suffering is the inevitable result. If we act with a pure mind, i.e. with the opposite qualities of contentment, compassion and wisdom, happiness will follow like a shadow.[47] We have to understand that pollution in the environment has been caused because there has been psychological pollution within ourselves. If we want a clean environment, we have to adopt a lifestyle that springs from a moral and spiritual dimension.

Buddhism offers humanity 'the middle way', a simple moderate lifestyle eschewing both extremes of self-deprivation and self-indulgence. Satisfaction of basic human necessities, reduction of wants to the minimum, frugality and contentment are its important characteristics. Every individual has to order their life on moral principles, exercise self-control in the enjoyment of the senses, discharge their duties in their various social roles, and behave with wisdom and self-awareness in all activities. It is only when each person adopts a simple moderate lifestyle that humanity as a whole will stop polluting the environment. This seems to be the only way of overcoming the present ecocrisis and the problem of alienation. With such a lifestyle, humanity will adopt a non-exploitative, non-aggressive, caring attitude towards nature. We can then live in harmony with nature, using its resources for the satisfaction of our basic needs. Just as the bee manufactures honey out of nectar, so we should be able to find happiness and fulfilment in life without harming the natural world in which we live.

Notes

All Pali texts referred to are editions of the Pali Text Society, London. Abbreviations used are as follows:

A	Anguttara Nikaya
D	Digha Nikaya
Dh	Dhammapada
Dh A	Dhammapada Atthakatha
J	Jataka
M	Majjhima Nikaya
S	Samyutta Nikaya
Sn	Sutta-nipata
Thag	Theragatha
Vin	Vinaya Pitaka

1 A IV, 100.
2 D III, 80.
3 D III, 71.
4 A I, 160.
5 Atthasalini, 854.
6 Dh A III, 854.
7 S I, 39.
8 Quoted in Vance Packard, *The Waste Makers*, Longmans, London, 1961, p. 195.
9 Dh, v. 204.
10 A IV, 2, 220, 229.
11 Dh A I, 20ff.
12 Dh A III, 129ff.
13 Vin II, 291.
14 A IV, 283.
15 D III, 188.
16 A III, 208.
17 Vin I, 137.
18 Vin IV, 125.
19 Vin III, 42.
20 Vin IV, 125.
21 D I, 4.
22 Sn, vv. 143–52.
23 J I, 191.
24 Dh A I, 58ff.
25 Vin II, 194f.
26 A I, 161.
27 J II, 423.
28 Petavatthu II, 9, 3.
29 S IV, 302; Dh A 1, 3.
30 D II, A.
31 S I, 33.
32 J I, 199f.
33 D I, 63.
34 M I 118; S IV, 373.
35 Vin IV, 205–6.
36 Robert Arvill, *Man and Environment*, Penguin, Harmondsworth, 1978, p. 118.
37 A III, 31.
38 M I, 457.
39 AS V, 15.
40 M I, 16–24.
41 S IV, 217, 293.
42 Sn, vv. 207–21; A I, 273.
43 M I, 126.
44 M I, 161.
45 Thag, vv. 1062–71.
46 Arvill, *Man and Environment*, p. 170.
47 Dh, vv. 1, 2.

3 THE SANDS OF THE GANGES

Notes towards a Buddhist ecological philosophy

Stephen Batchelor

Long before this age of astrophysics presented us with a universe in which the distances of stars were measured in millions of light years, Buddhists lived in a cosmos of similarly mind-numbing proportions. Theirs was a universe with no beginning or end and galaxies 'as numerous as the sands of the River Ganges' arose, existed and passed away across vast aeons of time.

Yet unlike scientists who peer at the data from radio telescopes and analyse the rocks of Mars in search of a glimmer of life, Buddhists saw the universe as teeming with diverse living beings, longing not just to be, but to be something, to have something, to feel something. This collective craving is seen as the reason for the existence and constant renewal of the universe. It translates itself into external environments—complexes of physical lifeforms; and into internal environments—complexes of thoughts, feelings and impulses with the tragic habit of grasping themselves as separate, solid and permanent selves.

Human existence is just one of six forms of life spread throughout the universe. Hedonistic and conceited gods, warring titans, animals, hungry ghosts and denizens of hell also inhabit the world systems, living out their own dramas and sufferings alongside human beings. Some are visible to the human eye, as with animals, but most are invisible, as with gods and ghosts. The presence of human beings is certainly not restricted to the planet Earth nor is the human species regarded as the best effort so far in an evolutionary unfolding of nature. Nonetheless, existence as a human is seen as an exceptional opportunity; for it is the kind of life most suited to finding out what is going on here.

While we in the West are inclined to think that we know what is going on, it is a fundamental principle of Buddhism that we do not. This is stated in the second of the Four Noble Truths which underpin

Buddhist philosophy. The first Noble Truth states that there is suffering. The second, that there is a cause for that suffering which is delusion. The third Noble Truth states that there is an end to the suffering (i.e. the delusion) and the fourth that there is a way or path to reach that end. When the Buddha spoke of enlightenment, nirvana, he was referring to the absolute understanding that ends our suffering.

THE FIRST AND SECOND NOBLE TRUTHS: SUFFERING AND ITS CAUSE

To remove suffering we must uproot its cause: delusion. And what is at the core of such delusion? In a word, separation. We each believe we are a solid and lasting self rather than a short-term bundle of thoughts, feelings and impulses. We feel ourselves to be separate selves in a separate world full of separate things. We feel separate from each other, separate from the environment that sustains us and separate from the things we use and enjoy. We fail to recognize them for what they are: part of us as we are of them, and the context in which we must painstakingly work out our salvation.

Delusion leads to all manner of problems. Our sense of separation reinforces the idea that we, or at least an important bit of ourselves, are somehow independent and unchanging. It lures us into believing that by accumulating enough agreeable pieces of reality—cars,

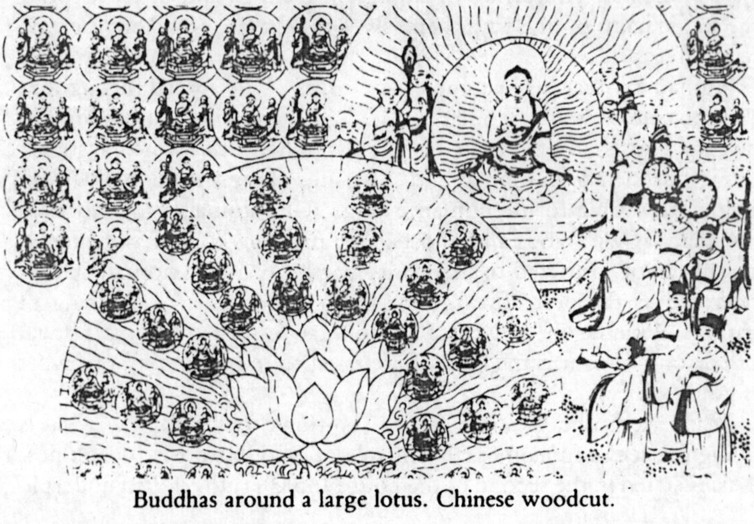

Buddhas around a large lotus. Chinese woodcut.

household appliances, clothes, hi-fi systems, fine art or whatever—we will accumulate a sense of well-being. It lures us into believing that the ability to control the world around us will one day cause a state of peace and happiness to arise within us.

This is not to say that many people would readily admit to holding such views. These views are far more insidious than that. Rationally, we all know that we are temporary creatures utterly dependent upon the ecosystem of our planet and finally destined to die. But we shouldn't fool ourselves into believing the views of our intellect necessarily bear any relation to our behaviour. The delusion of which Buddhism speaks holds sway over us in a much deeper way than mere ideas. We are in its grip almost physically, as though with our nerves, cells and chromosomes, it compels us to grasp hold of the world in a way that intellectually we would almost certainly reject.

So the ecological crisis we witness today is, from a Buddhist perspective, a rather predictable outcome of the kinds of deluded behaviour the Buddha described 2,500 years ago. Greed, hatred and stupidity, the three 'poisons' the Buddha spoke of, have now spilled beyond the confines of the human mind and village politics to poison quite literally the seas, the air and the earth itself. And the fire the Buddha spoke of as metaphorically engulfing the world and its inhabitants in flames is now horribly visible in nuclear explosions and smouldering rainforests, and psychologically apparent in the rampant consumerism of our times.

Perhaps we need these disasters to prompt us to consider more deeply what the Buddha was saying all along. For the ecological crisis is at root a spiritual crisis of self-centred greed, aided and abetted by ingenious technologies run amok.

THE THIRD AND FOURTH NOBLE TRUTHS: THE END OF DELUSION AND THE PATH TO IT

According to the Buddha, it is of little help to speculate about why our delusion is so, the fact is that it is so and something needs to be done about it. So, the spiritual challenge the Buddha set humanity was not simply to adopt a set of beliefs and thereby become good Buddhists. He spoke of 'practice' because the way we are requires urgent attention and action. It requires self-enquiry, discipline and insight sufficient to transform ourselves at our very core, so that we are no longer blinded by that tight addictive grip of delusion.

From the standpoint of the Buddha's enlightenment, the universe is a seamless, undivided whole. Only delusion makes it appear fragmented into an infinite number of separate pieces. The task of Buddhist practice is to recover this lost vision of wholeness and put the universe back together again. Practice, then, is healing. Not surprisingly, the Buddha is often referred to as 'the great physician', his teachings and their practice as 'medication', and those who help along the way as 'nurses'.

The healing process starts with putting our own life into order. We must consider the damage we are causing to the world around. Am I living in such a way that I am snuffing out other life? Am I taking for myself things that are not really mine? Am I betraying and hurting others through my sexual lusts? Am I leading a life of deceit and dishonesty, encouraging flattering fictions about myself and those I like and demeaning fictions about those I dislike? And am I undermining the clarity of my mind with drink and drugs so that I succumb to distorted views about myself and the world? All such behaviour only reinforces the sense of a separated and divided world and leads us further away from a vision of unity. If we are going to turn our lives around in a radical way, we have to begin by controlling the amount of actual harm we inflict upon ourselves, others and the world around us.

To think of spiritual development as an isolated activity that takes place in the privacy of our own soul merely reflects the very delusion of a separate ego that such activity is meant to overcome. Surely one of the most powerful modern delusions is the idea that we can do whatever we like so long as it does not cause any immediately noticeable harm. Yet the Buddha and others have bent over backwards trying to tell us that in fact *we do not know* what repercussions our behaviour might have; that ethical values cannot be determined by a deluded mind, but have to be based, until our delusions are dispelled, on the teaching of those with higher understanding.

An ethical life as prescribed by the Buddha's teachings is a 'practice' that provides protection against the urges of a deluded mind while also facilitating the healing of that mind. A healthy mind is one which has harnessed and channelled its energy away from cravings, excessive reactions, boredom and lethargy towards mental balance. It is on the basis of this balance that higher insights into the undivided nature of life can then arise. This is the point of deep personal transformation, of enlightenment.

Enlightenment, therefore, is not some mystical state where visions of unearthly bliss unfold, but a series of responses to the question: how am I to live in this world? Exactly how the insights of enlightenment surface in our minds is still a matter of debate among Buddhists. Some believe they are already present, waiting to be uncovered, whereas others insist that although the mind has the potential to realize them, we have to evolve over time to a point where they are finally attained. Regardless of whether we believe that insight will strike like a bolt from the blue or will gradually emerge over many years, we must be prepared for hard work.

NO SELF: THE INTERDEPENDENCE OF ALL

The Buddhist vision of reality is often spoken of in terms of absence. Insight is not only the discovery of something previously unknown and unsuspected but also the amazement that something you had always taken for granted has fallen away. You find that it is no longer necessary to uphold the fantasy of a solid, lasting self; reality works perfectly well without one and, in fact, this self has only ever managed to get in the way and cause trouble. The fear that denial of the self would give us no ground to stand on is realized to be in itself groundless, like the discovery we make as children when we find we can swim and are, at that moment, freed from the terror of drowning. Thus the instinctive insistence upon a separate self is seen to provide an utterly false sense of security; for in an undivided world everything miraculously supports everything else.

When the conviction that there is solid, enduring self co-existing with millions of solid, enduring others in a world of solid, enduring things, falls away, a universe of magically interrelated processes and events is revealed. That dreadful, alienating sense of separation dissolves, opening us to the freedom that is our birthright. Buddhist teachers often keep silent about the nature of this reality, knowing that for an unenlightened mind any descriptions would only tend to confuse and lead to speculation. Nonetheless, in some scriptures poetic imagery is employed to try and capture this sense of things as they are. One image used is that of Indra's Net, a vast grid of interconnected mirroring spheres, each one reflecting all the others (see p. 11). Uisang, a seventh-century Korean Buddhist monk, said:

> In one is all, in many is one.
> One is identical to all, many is identical to one.
> In one particle of dust are contained the ten directions.*
> And so it is with all particles of dust.

Twelve hundred years later, a similar insight was immortalized in the more familiar words of William Blake:

* See footnote on p. 11.

> To see a world in a grain of sand,
> And a Heaven in a wild flower,
> Hold infinity in the palm of your hand,
> And eternity in an hour!

Chinese Buddhist philosophers developed this vision of the world into the doctrine of 'the unimpeded interpenetration of all phenomena'. This doctrine stated that everything in the universe is literally dependent upon everything else, nothing stands alone, everything is linked together through time and space. Taken to its limit, the doctrine maintains that a speck of dust on Jupiter is intimately linked to a streetlamp in Tokyo, that a drop of water suspended from one leaf of a mahogany tree in a Burmese rainforest is united with the exhaust-fumes belching from a battered Chevrolet in Mexico City.

It is remarkable how little one has to tease out these ancient Buddhist doctrines to arrive at ecologically important statements. Long before environmental disasters brutally forced upon us an understanding of the interconnection of things, Buddhist teachers knew full well that such insight was crucial for the welfare of humanity. Moreover, these Indian and Chinese monks spoke of much more than the mere interconnection of the observable natural world: they included the vital role of the mind.

'The world', wrote the Indian Buddhist philosopher Vasubhandu in the fourth century, 'is created from intentions.' In other words, the environment we find ourselves in and the way we experience it are the consequences of how we have chosen and agreed to live. If our intentions are driven by self-centred greed and attachment, then that will determine the way we perceive the external environment, i.e. we will see it as a resource to be exploited to satisfy our desires and protect us against the things we fear. And since greed and attachment are short-sighted, mentally deadening and dehumanizing, the environment will reflect back those very qualities we inject into it. Decaying inner cities, gutted hillsides and polluted rivers are therefore the consequence of intentions of the human mind. To place responsibility for these things on the shoulders of industrialists and politicians, as we are prone to do, is just another knee-jerk reaction of a mind that insists on duality to make sense of life, in this case by dividing the world into 'innocent' and 'guilty'. Yet as long as we participate in the same delusions of separateness, then we too are responsible—by upholding instinctively a view of the world that allows such things to be possible.

Insight into the interconnection of life, however, is not just a personally satisfying solution to the problem of delusion. When we dissolve the rigid boundaries of the self, we inevitably reveal our connection with, and mutual dependence upon, other living beings. And when this insight breaks through in our hearts, it expresses itself as compassion and love.

The eighth-century Indian Buddhist poet Shantideva evoked this sense of universal sympathy with his image of life as a single organism, like a cosmic body. For just as the hand reaches out to a foot that is in pain, so does the enlightened person reach out in sympathy to those who are suffering. Insight into the interpenetration of all things transforms our immediate relationship with those around us, making it simply impossible to stand by with indifference and watch the world go up in flames. In a sense, the realization of the interdependence of life is a painful one. No longer can we remain comfortably insulated by the illusion of our separate selfhood. At this point, compassion stops being the deliberate doing of good, it becomes an instinctive urge. 'Although one acts in this way for others', remarks Shantideva, 'there is no sense of conceit or amazement. It is just like feeding oneself; one hopes for nothing in return.'

THE ECOLOGICAL CRISIS: IMPLEMENTING A BUDDHIST SOLUTION

For visions of doom and gloom, Buddhists have no need to refer to the messages of the modern prophets of impending ecological disaster. They have only to read their own texts to be told that within the huge time-cycles of the universe, humanity is currently embarked on a vicious downswing. The birth of the Buddha in this world was but a brief flash of cosmic illumination, in which the Dharma* was revealed to those with 'little dust on their eyes'. But it seems the Buddha did not expect the influence of his teaching to last very long. Estimates vary, but the general consensus among Buddhist traditions is that the Dharma would not remain for more than a few thousand years after the Buddha's death. Moreover, the fading of Buddhist doctrine is but one symptom of a period of moral degeneration, which is also characterized by a gradual shortening of human lifespan (in spans of time so vast that the recent increase of life expectancy in the affluent West could be interpreted as an aberrant blip), an increase in fatal diseases, the proliferation of weapons and, worst of all, a deepening of spiritual delusion.

If taken seriously, these predictions confirm the forecasts of scien-

* Dharma or Dhamma, the sacred law of the universe; the religious teachings; morality.

tists and others that we are living in a way that is both unsustainable and grossly irresponsible towards other forms of life as well as our own species. And if the world continues to be driven by the mounting forces of delusion rather than enlightenment, then what hope is there of implementing a Buddhist solution on a scale which could hold back the disasters looming towards us? In all honesty, are millions of human beings suddenly going to choose the Buddha's enlightenment instead of the pursuit of material affluence?

We cannot ignore these questions. If the Buddhist analysis of the ecological crisis is correct, then we are clearly going to have to do more than just switch to recycled envelopes and ozone-friendly hairspray to prevent the potential environmental catastrophe that a growing number of responsible voices are predicting. Yet to be realistic, we also have to accept that selfishness and greed are not going to vanish overnight. It would appear that the first step of a Buddhist solution—as 'skilful means'* at least—must be to explain how our present way of life is simply not in our own self-interest, let alone in the interest of millions of other beings and future generations.

The second step of a Buddhist solution would be to challenge the social structures which sustain and promote values that blind us to the ecologically destructive results of our actions. Two structures of particular importance would be education and economics. In the secular democracies of the West, both of these structures are based on belief in value-free (objective) knowledge, unlimited progress, and individual freedom. The combined effect of these beliefs is the rapid erosion of the moral values which still survive from our ancestral religions (compassion, generosity, self-control etc.). As a force of spiritual renewal, Buddhism would seek to inject into our social structures a fresh awareness of undisputed values—but without these depending on belief in God.

To be complete, however, both approaches (appealing to self-interest and renewing social values) must begin with an inner practice of self-transformation. Learning, reflection and meditation would uproot the tendencies of the mind which are destructive to both ourselves and our environment. We need to be encouraging within ourselves qualities such as simplicity, balance, compassion and understanding. We are each the starting point of a world-order based on these qualities. In such ways Buddhist practice would work inwardly at transforming the mind and outwardly at transforming the world.

* For 'skilful' see footnote on p. 5.

Moreover, there are two traditions in Mahayana* Buddhism which might prove particularly relevant to the ecological crisis. The first is the belief that all beings are enlightened if only they would realize it. For in spite of the scriptures' dire predictions of spiritual and moral decline, in spite of widespread ignorance and greed, and in spite of the breakdown of traditional values, delusion is not essential to life but accidental. In their innermost being, every creature is aglow with illumination. But like a sun obscured by a dark mass of cloud, their true nature is concealed. This is a doctrine of hope that can be raised to counterbalance the pessimistic leanings in Buddhism, which have prevailed for much of its history. When Zen Buddhists, for example, speak of sudden enlightenment, they mean that insight is something that can break into our lives at any time. Enlightenment is not a distant goal that we may reach after many aeons of effort but is already present here and now in everyone. To trust in the underlying presence of such enlightenment is a great strength in facing the calamities of life for it draws our attention to what is good in people rather than to the masks of delusion that hide their goodness.

The second tradition is that of the tantric doctrine of transformation. In tantric (or Vajrayana) Buddhism,** delusion is not thought of as something to be eliminated. On the contrary, it is seen as a particular energy pattern that is neither essentially good nor bad. When understood as energy patterns, even the most powerful delusions can be transformed into forces for enlightenment by transforming the pattern. Needless to say, the practice of tantra is a rigorous and demanding discipline which includes an element of danger. Yet it is often claimed by tantric teachers that the present 'degenerate' age is particularly suited to the Vajrayana, for the simple reason that the more intense the delusion, the more powerful is the energy available to be transformed into enlightenment.

We can also interpret the present delusive and dangerous situation as itself being a tantric teacher. For whatever our beliefs, the crisis *demands* that we act; it compels us to question ourselves in a way that constantly challenges us to transform our lives.

* For Mahayana see footnote *** on p. 8.

** Tantric Buddhism, also known as Vajrayana, 'Vehicle of the Diamond', is the 'Indestructible vehicle' for crossing the 'ocean of suffering' to enightenment. Tantric practice is based on the principle of transforming the impurities that defile the inherently pure soul. The tantric texts deal with the evocation of deities, the acquisition of magical power and the attainment of enlightenment through meditation, mantras (mystical chants), mudras (ritual movement/dance) and yoga. Tantra can only be received through the instruction of a guru. It is a form of Mahayana Buddhism.

SECTION B
THE PRACTICE

4 MAY A HUNDRED PLANTS GROW FROM ONE SEED

The ecological tradition of Ladakh meets the future

Helena Norberg-Hodge

The rough, winding road from Srinagar leads to Leh, capital of the old Kingdom of Ladakh. It climbs through the moss-green pine forests of Kashmir to the Zoji-la pass. The top of the pass forms a dramatic boundary. Ahead, in the parched rain shadow of the Himalayas, the earth is bare. In every direction are rugged mountains, merging into a vast plateau of crests. Above, snowy peaks reach towards a still blue sky; below, sheer walls of wine-red scree fall into stark, lunar valleys.

Yet, as your eyes begin to take in what lies before them, brilliant green oases come into focus, set like emeralds in a vast elephant-skin desert. Fields of barley, fringed with herbs and wild flowers, and the clear waters of glacial streams appear. Above, a cluster of houses, gleaming white, three floors tall and hung with finely carved wooden balconies; brightly coloured prayer-flags flutter on the roof tops. Higher still, on the mountain side, a monastery watches over the village below.

As you wander through the fields or follow the narrow paths that wind between the houses, smiling faces greet you. It seems impossible that people could prosper in such desolation, yet all the signs are that they do. Everything has been done with care: fields have been carved out of the valley walls and layered in immaculate terraces, one above the other; the crops are thick and strong and form such patterns that an artist might have sown their seeds.

Around each house, vegetable gardens and fruit trees are protected from the goats by a stone wall, on which cakes of dung to be used as fuel for the kitchen stove lie baking in the sun. On the flat roof animal fodder—alfalfa, hay and leaves of the wild iris—has been stacked in neat bundles for winter. Apricots left to dry on yak hair blankets and potted marigolds give a blaze of brilliant orange.

Lying in the shadow of the main Himalayan range, Ladakh is a high altitude desert criss-crossed by the gigantic ranges beyond. Life is dictated by the seasons; more so, perhaps, than in almost any other inhabited place on earth. Scorched by the sun in summer, the entire region freezes solid for eight months in the winter.

The great majority of Ladakhis are self-supporting farmers living in small settlements scattered in remote mountain valleys. The size of each village depends on the availability of water; generations ago, channels were built tapping the meltwater from the mountains above and bringing it down to the fields. The water is often channelled for several miles, across steep rock faces, stretching it as far as it will reach. At 10,000 feet and above, and with a growing season limited to four months of the year, the principal crop is barley. As elsewhere on the Tibetan plateau, the diet is based on its roasted flour, *ngamphe*. Most farmers also have some small fields of peas and a garden of turnips. At lower levels there are orchards of apricots and walnut trees. In the very highest settlements, where not even barley will grow, people depend largely on animal husbandry.

Like other traditional cultures which have grown up in hostile surroundings on the fringes of the habitable world, the Ladakhis exhibit an exceptional sensitivity in managing their environment and a keen awareness of their place in the greater natural order.

Culturally, Ladakh is almost pure Tibetan, and is in fact often referred to as 'Little Tibet'. Art, architecture, medicine and music all reflect this heritage. Tibetan Mahayana Buddhism is the predominant religion, the Dalai Lama is the overall spiritual leader, and for centuries Ladakhi monks studied in the great monasteries of central Tibet.

However, despite close cultural contact with Tibet, Ladakh was an independent kingdom from about AD 950. Leh itself was at one time a centre for trans-Asian trade and, therefore, was used to a cosmopolitan flow of foreign influences. Though it provided a source of luxury goods and commodities, this traffic did not affect the sedentary village economy. After 1834, when it was invaded by the Hindu Dogras, Ladakh fell under the rule of the Maharajah of Jammu and Kashmir. The region was never formally colonized and although sporadic wars and invasions over the last four centuries gave rise to some political and economic changes, they can scarcely have affected the traditional way of life.

Due to increasing tensions between India and Pakistan, the Chinese invasion of Tibet in the 1950s and their occupation of the Aksai Chin region in 1962, Ladakh became one of India's most vital strategic zones. Even then, the scarcity of resources and inhospitable climate served to protect the traditional society, as most military camps had no choice but to import the materials needed for their survival. It is only very recently, following the concerted efforts of central government to develop the region and open it to foreign tourists, that the traditional economy and culture of Ladakh has been seriously undermined.

I spent my first years in Ladakh analysing the language and collecting folk stories. I became fascinated by the people—by their values and the way they saw the world. Why were they always smiling? And how did they support themselves in relative comfort in such a hostile environment?

When I arrived as one of the first outsiders in several decades, Ladakh was still essentially unaffected by the West, but change came swiftly. The collision between the two cultures has been particularly dramatic, providing stark and vivid contrasts. Over the past sixteen years I have had the privilege of experiencing another saner way of life, and at the same time witnessing the impact of the modern world on that culture. It has been a rare opportunity—and a tremendous inspiration—to compare our socioeconomic system with another more fundamental and harmonious pattern of existence; a pattern based on co-evolution between human beings and the earth.

LIVING WITHIN THE SEASONAL RHYTHM

Agriculture in Ladakh is closely co-ordinated with the seasonal movements of the sun and the stars. The timing of the cycle varies with altitude: every village will determine its own calendar from the shadows cast by a solar obelisk or *nyitho* sited nearby. When the sun casts its shadow in the right place for the sowing season to begin, the astrologer is consulted to choose an auspicious day, hopefully one on which the elements of earth and water will be matched. Someone whose signs he deems favourable is chosen to sow the first seed. Next the spirits of earth and water, the *lu* (or *nagas*) and *sadak*, must be pacified; the worms of the soil, the fish of the streams, the soul of the land. Any interference with the soil, digging earth, or breaking stones, is liable to upset or anger them. Before sowing, a feast is prepared in their honour. For an entire day, a group of monks recites prayers, and no one eats meat or drinks *chang* (barley beer). In a cluster of trees at the edge of the village, where a small mound of clay bricks has been built for the spirits, milk is offered. As the sun sets, other offerings are thrown into the streams.

Manure has been brought with donkeys and heaped by the fields: at dawn the women quickly spread it in the furrow. At sunrise, the whole family gathers, men carry the wooden plough, and children lead the *dzo* (a cross betwen a yak and a cow) to be yoked. They set to work in a festive atmosphere, laughter and song drift back and forth across the fields, while a robed monk sustains a solemn chant. The *dzo* pull the plough at a dignified and unhurried pace. Behind, the sower throws the seeds and sings:

Manjusri embodiment of Wisdom, Hark!
The gods, the nagas, owner spirits of the Mother Earth, Hark!
May a hundred plants grow from one seed!
May a thousand grow from two seeds!
May all the grains be twins!
Please give enough that we may worship the Buddhas and bodhisattvas,
That we may support the Sangha and give to the poor!

Tibetan Thangka painting.

My first experiences of traditional life were in the village of Hemis Shukpachan ('place of juniper trees'). My friend Sonam, a government clerk in Leh, was returning to his family house at the time of *Skangsol* or Harvest Festival, and had invited me along. I remember waking to the fragrant smell of burning juniper. Sonam's Uncle Phuntsog was walking from room to room with an incense burner, a daily ritual of purification. I walked out onto the balcony—whole families were working in the fields, some cutting, some stacking, others winnowing the crop. Each activity had its own song. The harvest lay in golden sheaves, hundreds to a field, hardly allowing the bare earth to show through. Clear light bathed the valley with an intense brilliance. No ugly geometry had been imposed on this land, no repetitive lines. Everything was easy to the eye, calming to the soul.

Meanwhile in the family shrine the monks were performing ceremonies for Skangsol. They had made pyramids of barley dough *storma* decorated with butter and flower petals as offerings to the five *Dharmapalas* or protective deities. Prayers were offered for the happiness and prosperity, not only of this family or village, but for every sentient being in the universe. At evening, people gathered to sing, drink and dance. A butter lamp was lit in the kitchen, and garlands of wheat, barley and peas strewn around the wooden pillars.

The crop is threshed on a large circle of packed earth by a team of animals hitched to a central pole. Winnowing is especially graceful: two people facing each other will scoop the crop into the air with wooden forks in easy rhythm. They whistle as they work, inviting the wind:

> Oh pure Goddess of the wind!
> Oh beautiful Goddess of the winds!
> Carry away the chaff!
> Ongsla Skyet!
> When there is no human help
> May the Gods help us!
> Oh beautiful Goddess!
> Ongsla Skyet!

The grain is then sifted. Before it is put into sacks, a little figure or painting of a deity is ceremoniously placed on top of the pile, to bless the harvest.

I was beginning to experience the 'wholeness' of this way of life. For the Ladakhis there are no great distinctions or separations between work and festivity, between human spirituality and attendance to the natural environment. All one's actions are integrated and given meaning in the cycle of existence.

LIVING WITHIN NATURE'S MEANS

Soon after arriving in Ladakh, I was washing clothes in a stream. Just as I was plunging a dirty dress into the water, a little girl came by, perhaps seven years old and from a village upstream.

'You can't put your clothes in that water', she said shyly, 'people down there have to drink it.' She pointed to some houses at least a mile further downstream. 'You can use that one over there. That's just for irrigation.'

I was learning how the Ladakhis manage their difficult environment. I was also learning the meaning of 'frugality'. In the West, frugality conjured up images of old aunts and padlocked pantries. In Ladakh, where frugality is fundamental to people's prosperity, it has quite a different meaning. Being careful with limited resources is not miserly—rather it is frugality in the original sense of 'fruitfulness', getting more out of little. Where we would discard something as worn out or beyond repair or useless, the Ladakhis will always find further employment for it. What cannot be eaten can be fed to the animals, what cannot be used as fuel can help to fertilize the fields. A woollen robe will be tirelessly repatched until, when no amount of stitching can redeem it, it will be packed with mud and used to shore up an irrigation channel. Sonam's grandmother would shape the crushed apricot kernels, from which oil had already been extracted, into small cups which, after hardening in the sun, would serve to turn her spindles. The wild shrubs and bushes that we would regard as weeds provide medicine, incense, basket fibre and so on. Most notably human waste is composted to be used as fertilizer, and earth closets are found in all traditional houses.

In such ways, the Ladakhis manage to recycle everything—there is literally no waste. With only scarce resources at their disposal, farmers have achieved almost complete self-reliance, dependent on the outside world only for salt, tea, and a few luxuries. The contrast between this frugal management of resources and our modern high-consumption, high-waste economy is extreme. However there is a striking similarity between these age-old practices and the notions of 'ecology' we are now developing.

Some years later, at a meeting I arranged to discuss the changes in modern Ladakh, the villagers were debating whether the young generation were losing respect for traditional practices and values. Some youths denied it, but an old man interrupted:

> Sure! Ask them to saddle a horse and they put it on backwards! They

> buy expensive rubber boots that fall apart before you reach the top of the pass. We wore shoes we made ourselves that were warm and comfortable. We stood on our own two legs and knew how to make use of everything around us. That's what you mean by 'ecology', isn't it?

Back in Hemis, Uncle Phuntsog could often be seen weaving in the shadow of a large walnut tree. Surrounded by a group of young helpers working the pedals of the loom, he told stories punctuated with song, and was constantly plied with tea and chang from the house. Despite many distractions, and without ever seeming to, he worked remarkably fast, completing a full length of cloth in one day.

The Ladakhis get wool either from their own animals or by trading surplus grain. They wash it, spin it, weave it, dye it, and sew it themselves. Spinning is a constant activity: men and women alike will spin as they walk with loads on their backs or sit by the stove in the evening. It appears to be a means of relaxation, almost a form of meditation.

A new house is never built without concern for the sadak or earth spirits. A *lama* (religious teacher) will come to bless the land. He then uses a brass mirror to reflect all the surroundings, capturing the sadak to protect them from harm during construction. The mirror is placed in a box where it remains until the building is complete. In a final ceremony, the lama will open the box and set the spirits free.

Ladakhi houses are large structures, two or three floors high. The whitewashed walls taper slightly inward to the flat roof giving an impression of grace despite the massive proportions. They offer a comfortable refuge from the vast and unsheltered expanse outside and are usually built by the family themselves. Apart from the stone foundations the main material is mud bricks. The walls are plastered and whitewashed and poplar beams support the roof. Only the aesthetic details, carved lintels and balconies, require specialized skills.

So it is that the Ladakhis work their own lands for their food and produce their own homes and clothing all from materials available around them in designs that are both functional and attractive. Work is varied and shared and, apart from such vocations as metal work and artisanship, its techniques are within everyone's experience and comprehension. Living and working together in a society on this scale, where collective duties and decision-making are usually shared, each member has an overview of the structures and networks of which they are a part. When individuals can experience themselves as part of the community and see the effects of their actions on the whole picture, it is easier both to feel secure and to take responsibility for their own lives.

Altogether, the Ladakhis use only the simplest technologies to provide their basic needs: the plough, the loom and a water powered mill to grind grain. The tasks for which our society would employ heavy machinery are handled with team work or animal labour. While the industrial worker gets the job done fast and becomes the appendage of the machine, the Ladakhis are unhurried and remain fully in control of the few machines they employ. They know how to make and repair them and the loom or water mill is not so noisy as to prevent them from singing or chatting as they work.

And of course they have the time. Ladakhi measurements of time are broad and generous, rather than precise and universal: they include such words as *nyi-tse* or 'sun on the mountain peaks' and *chipe-chirrit*, 'bird song', the time just before dawn. Even during the harvest season when there are long hours of hard work, it is done at a relaxed pace that allows an eighty-year-old as well as a young child to join in and help. And the Ladakhis only work hard for four months of the year. Most of the long winter is spent at the monastery festivals and social gatherings, and sitting by the fire, telling and listening to a rich heritage of stories.

LIVING RELIGION

As in all Tibetan Buddhist societies, the highly developed monastic system is the most prominent feature of Ladakhi culture and social organization. The massive presence and role of the monasteries reflects the society's priorities—virtually all the resources and energies of the people beyond the satisfaction of their daily needs is sunk into these storehouses of learning and wisdom. It is a culture in which the higher values of Buddhism are prized above all.

This system often evokes for us the image of an idle priestly caste exploiting the peasants and conjuring demons to terrify them into submission. The fact is, however, that the relationship between monastery and village in Ladakh is one of mutual support and benefit. Monasteries often own a good deal of land which is worked by the villagers, and monks are provided for by the whole community in exchange for a variety of social, religious, and spiritual services.

On the economic level, the monastery acts as a grain bank and works hand in hand with the polyandry system to balance population and land holdings. At least one member of each family will join the monastery (or nunnery) and it is an option open to all, male or female, young or old, as an alternative to the life of a married householder. Although the Ladakhis have relatively well developed secular equivalents, the arts of painting, sculpture, printing, music and dance, and the sciences of health care and astrology are otherwise the preserve of the monasteries, in addition to the vast array of spiritual knowledge, technique and ritual that is the heritage of Vajrayana Buddhism.* The monks' duties will often take them to people's homes to perform important rituals and read the scriptures. The process of give and take between the monastery and village sustains a rich cultural and spiritual life in which the whole community is involved and benefited.

From daily prayers to annual festivals the entire calendar is shaped by religious beliefs and practices. Every week of the Tibetan lunar month has some significance. The tenth day belongs to Guru Rinpoche, the great tantric apostle; the fifteenth marks the full moon, the day when Buddha was conceived, attained enlightenment, and passed beyond. Villagers gather at each other's houses to read the scriptures together; they will print fresh prayer-flags on cloth of the five auspicious colours and there will be eating and drinking. For *Nyeness* in the first month of the year, people assemble in the monastery to fast and meditate together.

For the monastery festivals, villagers congregate in hundreds and thousands to watch *cham* dances in which the basic teachings of the Tibetan tradition are enacted in theatrical form. The sounds of horns and drums blend with the chanting of mantras and laughter. The monks dance in splendidly colourful costumes and masks representing various figures of the Tibetan pantheon, all of which have a deeper symbolic meaning. The performance climaxes in the ceremonial destruction of the ego, the enemy of spiritual liberation. This event combines celebration, entertainment and instruction, in which religion mingles with folk tales and pantomime: the sacred embraces all aspects of life, utterly. But despite the attention they attract, these rituals are not of central importance. A senior lama once told me: 'As long as there is ignorance, there is a need for ritual. It is a ladder which may be discarded once we have attained a certain level of spiritual development.'

Beyond the monastery, Buddhist symbols and objects of devotion are found everywhere in people's houses and scattered across the landscape. *Chotens* (stupas**) made of whitewashed mud and stone surround every village and monastery and adorn even the remotest

* For Vajrayana Buddhism see footnote ** on p. 39.

** Round steeple-like monuments, traditionally for the ashes of monks or storing holy relics.

valleys and passes. They are often crowned by a half moon cradling the sun, a symbol of the end of duality, the oneness of life and the inextricable relations that connect all beings and phenomena. The *spallbi* or endless 'knot of life' has a similar significance. To walk around the choten clockwise is itself an act of devotion, much as turning the ever present prayer wheels. The traveller in Ladakh also frequently encounters prayer walls: stones carved with prayers and mantras are left here as offerings and accumulate through generations into impressive monuments.

Almost every house has its own, often splendid, shrine room tended by the family. A large prayer-flag pole, *tarchen*, outside the house indicates that it possesses the full sixteen volumes of the basic Mahayana* texts, the Prajnaparamita ('Perfection of Wisdom') Sutras. Evidently there are many devoted and capable lay practitioners; and religious activity, while dominated by the monasteries, is nevertheless shared by the community.

Thus in the course of everyday life in Ladakh the people are constantly reminded of the higher truths of Buddhist philosophy: the interdependence of all things, the ultimate emptiness of apparently separate, independent beings. These sophisticated theories of reality and the lofty values of the scriptures are not confined only to the lamas who are versed in them, but, to a great extent, shared by all.

For me the most profound expression of Buddhism lies in these more subtle values and attitudes among the people. Although deep meditation is rarely practised outside the monastic community, people spend significant periods of time in a semi-meditative state. Older people in particular recite prayers and mantras as they walk and as they work—even in the middle of conversation. Recent research in the West suggests that during meditation a person enters the state of mind that perceives in wholes and patterns rather than by isolating and itemizing things. This may explain the holistic or contextual world view characteristic of the Ladakhis, even those who have little knowledge of the teachings.

It could be argued that there are traces of Buddhist awareness even in the Ladakhi language. There is a greater emphasis on relativity than in any Western language I know. The language obliges you to express the context of what you are trying to say. Most strikingly the verb 'to be' has more than twenty variations, depending on, for example, the relative intimacy of both speaker and listener with the subject matter and the relative certainty with which something is stated. How I say 'it is milk' will depend on whether it is my milk (if

* For Mahayana see footnote *** on p. 8.

not, then whose), whether I can see the milk, and so on. And if I ask someone, for instance, 'is it a big house?' he or she is likely to answer 'it seemed big to me'.

The same relativity or contextuality occurs on a conceptual level. The Ladakhis do not share our enthusiasm for categorizing and compartmentalizing the world. It is as if the awareness of ultimate 'emptiness' has undermined the commonplace approach to things as separate, definite entities. In their attitude to matter and to time there is a reluctance to make concrete affirmations or certain distinctions. For example, good and bad, fast and slow, here and there are not sharply differing qualities but aspects of a continuum, a matter of degree. In the same way, Ladakhis will not think in terms of a fundamental opposition, for instance, between Mind and Body or Reason and Intuition. They experience the world through their *semba*, best translated as a cross between 'mind' and 'heart', which reflects the Buddhist insistence that wisdom and compassion are inseparable.

Also reflecting the Buddhist view of reality, the Ladakhi sense of self is based on a complex web of interconnection and constant change, rather than a notion of static isolated individuality. They tend to be more open to the uniqueness of a moment or a situation, and they have little difficulty in letting go and feeling at one with themselves and their surroundings. Their sense of self is extensive and inclusive rather than retreating behind boundaries of fear and self-protection; and it gives rise to a gentle yet unshakeable confidence and dignity.

Ladakhi attitudes to life and death are imbued with an intuitive understanding of impermanence, from which a lack of attachment grows naturally. I have been struck by the ability of my Ladakhi friends to welcome things as they are, to feel happy regardless of circumstances. Whether they are coming or going, in rain or shine, is really not so important, since their contentedness and peace of mind are not dependent on outer things. They may be unhappy to see a friend leave or to lose something valuable, but not *that* unhappy.

The same sincere acceptance of worldly impermanence is apparent in their acceptance of death. Life and death are known to be a single process of constant renewal and return. Ladakhis do value life highly, but they see the present life as only one of many, and will not cling to it in dread of death. This amounts to a profound understanding and acceptance of the human reality, which seems to be as much a source of joy as of sorrow. Again there is a difference in degree: of course there is misery and a sense of profound loss when a loved one passes away, but not the same desperate finality that we often feel.

LIVING WITH THE WEST

Rapid change descended upon Ladakh in the years following 1974, when the central government threw the region open to tourism and initiated the 'development' process in earnest. As everywhere else in the world, 'development' means Western-style modernization. It came primarily in the form of motor roads and energy production, followed by Western medicine and education. The 'formal sector', a law court, police department, banks, government administration, radio and television, became established in Leh alongside the military sector. The money economy, virtually non-existent outside of Leh until now, was stimulated at every level. Truck traffic increased in leaps and bounds and air pollution with it, bearing goods like wheat and rice, coke and firewood from Kashmir and Punjab. Population levels soared and modern housing construction went into an upward spiral around Leh. Waste, water pollution and its related diseases, all hitherto unknown, became normal. Electricity poles replaced trees, flaking paint, rusting metal, broken glass, plastic rubbish are now part of the scenery; billboards advertise cigarettes and powdered milk.

Tourists have arrived in their thousands during the summer months, and with them a booming economy of hotels, restaurants, guides and souvenirs. More important still has been the effect on people's minds. The tourists are creatures from another world who apparently have special powers, constant leisure and inexhaustible wealth. Many would spend in one day the equivalent of a local family's annual income. I realized that our culture looks infinitely more successful on the outside than we experience it to be on the inside. For the young Ladakhis, traditional culture seemed primitive, silly and inefficient when contrasted with this life of glamour and wealth, and many felt stupid and ashamed. The tourists themselves only saw the apparent poverty and backwardness of Ladakh; its social and spiritual wealth was invisible to them.

Switching to the cash economy has made Ladakhis dependent on the faraway forces of the international system, vulnerable to inflation and unemployment, where previously they had been self-sufficient and self-determining. Until now, these developments have been mainly confined to Leh and its immediate environs, but as the modernization process takes hold, one can see it expanding further into the village economy. Greed has multiplied and disparities of wealth have deepened where previously they were insignificant. As land and labour assumed monetary values and imported materials are subsidized, it became 'uneconomic' to grow your own food, a hitherto unimaginable concept. Many are obliged to leave the villages and earn money in the city. Cement replaces mud, synthetics replace wool, rice is preferred to barley, plastic to brass. The wisdom of the lama is less respected than the knowledge of the engineer.

Community ties and co-operation have gone into decline: people cease to feel that they need each other when social integration becomes mediated by anonymous bureaucracies, money and technology. Education joins up with advertising media and tourism to emphasize conformity to the norms of the global monoculture, and reinforce the negative image of traditional culture as inferior. When school children were asked to imagine the future of Ladakh, one little girl wrote: 'Before 1974, Ladakh was not known to the world. People were uncivilized. There was a smile on every face. They don't need money. Whatever they had was enough for them.' Another child wrote: 'In these days, we find that maximum people and persons didn't wear our own dress, like feeling disgrace.' And Dolam, aged 8, wrote: '. . . there will be nothing left which can prove the culture of Ladakh.'

As the disintegration of this gentle and sophisticated culture began to unfold around me, I came to realize that the Ladakhis had no information about the negative aspects of 'development', the social and environmental costs now well known in the West. Still less were they aware that many Westerners had become disillusioned with the fearful destructiveness and spiritual poverty of industrial society and were exploring more natural alternatives. Organic agriculture, soft energy systems, herbal medicines and even the Buddhist Dharma—all very much a part of traditional Ladakh—are the most modern developments in the most 'developed' countries.

I became involved in exchanging information about global issues and the search for sustainable alternatives to conventional development both in Ladakh and the West. The initial step was to introduce solar power (for which there is an enormous potential on the Tibetan plateau) as an alternative to imported solid fuels which are costly and create pollution. We successfully installed 'Trombe Wall' heating insulation in traditional houses. We organized plays and radio programmes to spread our message. By 1980 these activities had grown into a small international organization called the Ladakh Project.

In 1983 we helped to launch the Ladakh Ecological Development Group, together with a forum of eminent Ladakhis, educated and dedicated people, concerned to sustain their traditional culture and promote 'development without destruction'.

We have introduced solar ovens and water heaters, greenhouses to extend the growing season, hydraulic ram pumps to extend water supply using the power of gravity rather than petrol, and micro-hydro-installations to provide electricity. Our work aims to involve and train villagers to use technologies which augment rather than undermine traditional lifestyles and values. This has been accompanied by educational and handcrafts programmes. Over the years it has become the most influential non-governmental organization in the region and has won widespread interest and support.

Ladakh cannot and perhaps should not return to the past, but within the context of rapid modernization, our efforts have been and continue to be an attempt to create an alternative path into the future.

For further information

The Ladakh Project, 21 Victoria Square, Clifton, Bristol BS8 4ES, UK

From the ten *Gentling the Bull* ('Oxherding') pictures. Zen.

5 CIRCLING THE MOUNTAIN

Observations on the Japanese way of life

W. S. Yokoyama

INTRODUCING THE PROBLEM

At a recent symposium, a number of papers were presented discussing the issue of Buddhism and ecology from various points of view. On the panel was an orange-robed Buddhist monk from south-east Asia. Though attentive to the proceedings, he seemed content not to utter a single word. When at last asked by the chairperson whether he had any comments to make, the monk rose and said: 'All these papers are very good, but what do they have to do with the self?'

Perhaps what our friend in the orange robes is trying to tell us is that the real solution begins with us. If we think that the problem lies 'out there', then we might also imagine that any solution, such as one Buddhism might offer, lies 'out there'; all we have to do is to make our lives 'fit' that external ideal. But like the confused man dashing madly about in search of his head, no real solution is forthcoming until we come to our senses and turn our attention in the right direction.

Think for a moment what it is that we are trying to protect the environment from? Is it to protect it from those others, from 'them'? Buddhism strongly suggests that we must first protect the environment from ourselves. These days, the face of the adversary is complex; it is not just the large multinational corporations that are destroying the rainforests in some distant country. If we are to engage the enemy at its ultimate source, we have to reckon with ourselves and the lives we are leading. For though the hand that fells the tree may not be the same as the gilded one holding the toothpick, they work in league to feed the vanity of the self.

Some scientists regard the development of the mind and its sense of 'self' as an important step in evolution. Over the vast expanse of time since life first began, the human mind has been the instrument that has enabled our particular species to survive. But the problem with the mind is that, in dealing with the environment, it can only do so by filtering out reality. Let us say we are searching for a pencil in our room. Imagine if every object that fell into our line of vision was brought under our mental scrutiny. This would create havoc. Instead, our mental processing acts like a filter to select what is and is not the object of our search. Much of our waking time is spent with this mental processing 'on', during which time our mind constantly deliberates over one thing and another.

Because the mind sees only what it needs, or thinks it needs, to survive, we end up moving through life at a distance from reality. Not only do we fail to see things we consider irrelevant to us, but even what we do see is interpreted to fit our beliefs and expectations of the world. The gap between 'what is' and 'what we think is' is often explained using the analogy of the person walking in the forest who is alarmed by a rope because they think it is a snake.

While citing the limitations of the human mind as the source of all delusion and therefore all suffering, Buddhism also recognizes humanity's potential to overcome them. Transforming evil into good, as Buddhism boldly claims to do, means that, by taking the proper measures, it is possible to re-route a person from harmful ways of thinking and behaving toward beneficial ones. Society seeks to attain this goal by educating its youth. So, by the time we are young adults we are expected to know enough to show good judgement. Buddhism seeks to attain similar goals by meditation, a long process during which the judgement of the seeker matures.

THE BUDDHIST SOLUTION

Sitting perfectly still may seem an odd way to go about solving a problem. Ordinarily, we think we should 'do something' when we have a problem, not merely sit around and 'do nothing'. But there is more to meditation than meets the eye. Meditation as the Buddhists have devised is a highly practical way of achieving a state of calm in which the seeker perceives reality in perfect clarity.

At every stage when we meditate we must actively 'do something'. We must first concentrate on a certain object, for example, the breath. We must be aware of the movements of the breath and the sensations that arise from it. This requires constant attention. We must guard against sleepiness or mental agitation. If we are successful in this, we develop an inner state of deep calm and clarity. This does not mean that our mind goes blank like a switched-off television screen. Rather, our mind becomes so aware that the field of mental processing is removed and our mind is able to perceive reality as never before.

The model of Buddhist meditation presented here is a highly practical one that rejects the idea of meditation as a mindless mystical trance. On the contrary, the seeker abides in a state of calm in which mental focus and acute awareness play an important role in dissolving the dualistic thought processes of 'this or that' and of reducing mental dullness and agitation. This model is the cornerstone of the Buddhist way of life that has assumed various forms throughout its history. In the next section, we will look at a few examples of how this model is actually applied in Buddhist life in Japan today.

A WAY OF LIFE: THREE BUDDHIST TRADITIONS OF JAPAN

The Tendai monastery

It is early morning on Mount Hiei, north-east of Kyoto. We are among a group of visitors experiencing monastic life at the Tendai Centre. The cool morning air mixed with a thin fragrance of incense makes the long wooden meditation hall we are sitting in appear a world apart from the stifling tropical heat of the city below. Sitting in meditation on the round cushions in two rows, the only sounds we can hear are the rise and fall of our own breath and the occasional warble of a nightingale outside. Somewhere behind us a tiny handbell is struck. We are told to keep our mind focused on the sound as it gradually fades . . .

Perched on my cushion in the second row, I find it hard to keep my thoughts from wandering. Still revelling in the cool morning air, I wonder why more people don't come up here to sit in meditation. Then I recall what the priest had said about how cold the winter is up here and the other practices.

In another hall a short distance away, a young priest is doing the constant walking meditation. In this practice, the seeker is confined to a hall for 90 days during which time he must walk clockwise round the Buddha image without stopping. As he circles, he must also hold the image of the Buddha in his mind and recite the Buddha's name. Austere practices such as these are thought to trigger a spiritual experience, but the goal is not easily achieved. The mind plays tricks on the practitioner, especially in the dead of night when the hall is pitch-black. Demons and Buddhas appear from out of dark corners to frighten or lure the weary walker from his practice. He is instructed to ignore the figments of his imagination. Whatever happens, he must continue to walk.

Our meditation session lasts only 30 minutes. Despite the briefing on how to meditate the night before, I am afraid my efforts to put it into practice were rather puny. As the others begin to file out of the hall, I am left behind, my feet having gone numb from sitting. In an opening in the forest, the other members of the group are already lined up for morning exercise. Some of them are bank employees; others appear to be military men and college students. The programme is open to anyone who wishes to participate.

Most of the people in our group are no doubt already familiar with the important role the Tendai school played in Japanese history. It was partly through the Buddhist monasteries that the advanced culture of China and Korea was introduced to Japan, then an insignificant scattering of islands off the eastern coast of the Asian continent. Later, the Tendai school, with its wide variety of practices, became the source of inspiration for many of the popular Buddhist schools that developed within Japan. It did this, not by imposing its religion on the people, but by assimilating the positive values basic to the Japanese way of life such as love of the land and mountain worship. In this way, the Buddhist monastic centres such as Mount Hiei were able to re-direct the basic instincts of the people toward improving their country, at the same time preparing the soil for the religious culture of today. To the Japanese, then, Mount Hiei is not just a monastery separate from the world of daily life below but an integral part of their own religious and cultural heritage.

As we start to ascend the stone steps back to our lodgings, a figure in white robes carrying a staff dramatically appears over the ridge. It is a Tendai practitioner, well into his sixties, who is engaged in the thousand-day pilgrimage. The pilgrimage involves walking a fixed path around Mount Hiei for a thousand days, an ordeal which few practitioners have ever been permitted to do. Everyone bows reverentially as the figure approaches our group. As he draws closer, we see he has a radiant face under his unusual boat-like hat. Appearing to be in no particular hurry, though no doubt he had a good distance to go, he stops to talk to us for a few minutes.

The story of this Tendai practitioner is that he had been in business once. This was during the wartime when it was hard to make ends meet as an honest man. Later, when his wife died, he wanted to make good his promise to become a good man. Leaving that world behind, he became a monk on Mount Hiei. After years of practice, he was finally allowed by his teacher to engage in the thousand-day

From the ten *Gentling the Bull* ('Oxherding') pictures. Zen.

pilgrimage. Walking daily, whether through winter snow or summer heat, he ultimately reached a point where he entered a state of 'oneness with nature', attaining what Buddhists regard as the ultimate awakening in this life.

The Tendai practices date back centuries and, barring unforeseen disasters, will continue for centuries more, but it would be difficult to argue that they serve as a model for our age. There are few individuals who would want to or are suited to undertake the rigours of practice being done on Mount Hiei. For most people, myself included, a single mosquito bite or stubbed toe is enough to put a damper on our enthusiasm for monastic life. As such, deep spiritual experience is closed to ordinary people like us.

However, there is something suggestive about the practitioner's experience of 'becoming one with nature' that we must not overlook. With the industrialization of society and the concentration of large populations in metropolitan areas, modern life has grown more and more distant from nature. The thousand-day pilgrimage in which the practitioner contemplates the mountain reverses this trend. As the practice deepens s/he enters into communion with the delightful beauty of the world, its rocks, trees and flowers.

The practitioners' experience is not merely an aesthetic one, however, for it returns them to their original point of balance within the natural world, beyond the world of delusions created by the human mind. Whether or not we can do the arduous practices of the Tendai school, it is no less important for us to restore this balance to our way of life.

The Zen temple

The Zen master was a powerful speaker despite his years. His Sunday morning talks at the Zen temple always attract large numbers of people. Today was no different; the room was filled to capacity. The audience is mostly neatly dressed older folks who, in another setting, would be devout church-goers. There are also a few Westerners in attendance, expatriates who had come to Kyoto to do, among other things, Zen practice.

On this particular morning, the Zen master has us thinking. Acting as devil's advocate, he makes a few sharp comments about recent advances in the medical world in Japan which received favourable press coverage in the Japanese newspapers. Why is so much effort being put into these operations, he asks, when there are so many people who are already starving in this world? What is it that we fear? If it is dying that we fear, should we not instead learn how to care for the stricken child, rather than think how to cure the child's condition? Further into his talk, he remarks, 'If this trend continues, what will happen to us? Are we to buy our air in bottles, just as we buy bottled water?'

The audience is extremely polite, and it is difficult to tell whether they agree with the Zen master's views or not. If nothing else, it certainly prompts us to think more critically. Because of the high literacy level in Japan, it is easier for the media to herd people by prodding them with one news-breaking story after another. Are we to accept everything as reported in the press as gospel truth? The thought-provoking Zen master wants us to get down to real issues.

Two centuries ago, Zen monasteries such as this one occupied the best parts of town. This situation came about because the Zen school was patronized exclusively by the warrior class which then controlled the feudal government. The arts of war not being in demand in this peaceful age, the warriors-turned-bureaucrats shifted their energies from the martial arts to the practice of Zen-inspired arts such as the tea ceremony. This accounts for the marked influence of Zen Buddhism on Japanese culture.

Also in those days, the stratification of Japanese society into castes meant that ordinary people had no opportunity to come into contact with the Zen world; for instance, warriors and farmers simply did not mix. With the political changes in the last century, however, the Zen temples lost the patronage of the rulers. This change also opened up the Zen school to ordinary people, such as those who gathered here for the Sunday morning talks.

Aside from this polite audience of listeners, there is another more

Two pictures by Giei Satō.

serious group the Zen master attends, the students in training. Some of the students are leading a monastic life, others the life of laymen. The Zen master's role is to see that they are moving in the right direction. His whole life is devoted to their training.

Zen students are trained by the unique *koan* system. The Zen master assigns a koan to the student whose mission is to solve it. A koan is a sort of question or puzzling statement that the Zen student has to work out. For example, 'A disciple asked Joshu, "Why did Bodhidhamma (the first Zen Patriarch) come from the West?" Joshu said: "The cypress in the courtyard".' At regular intervals the student goes to the Zen master to give his answer or explanation. The answer, if it can be called that, is not one that can come from 'outside'; it can only come from the hidden resources of the human heart. Developing these resources in the student is the task of the Zen master.

Since the Zen student must concentrate on the koan until it is solved, the koan can be regarded as an aid to meditation. It is a meditation aimed at awakening the dynamic functioning of the mind. The Zen master uses the koan to test the depth of the student's understanding, just as one might test the depth of a well by dropping a pebble into it.

Attaining awakening is important, especially in the training programme for future Zen masters. Not all of us are suited for the rigours of such training, however. In the future, it is important to think of what Zen means in terms of the ordinary layperson.

Another seeker continues the Zen life. He has already 'done his time' in the Zen monastery which is a short distance from where he now lives. His lifestyle remains a simple but relatively disciplined one. Rising early each morning, one of his first chores is to tend the vegetable garden he has planted in a sunny corner of the temple grounds. The garden is actually quite a sizable one and occupies him year round. But the tomatoes, eggplants, lettuce and so on are enough for his needs as well as those living in the temple. In one corner of the garden he maintains a compost pile. Next to it he splits the kindling and stacks it neatly on the woodpile for the people in the temple to use for the fire that heats the bath. His garden duties done, he sets off by bicycle for the Zen college across town where he teaches. Asked where this kind of life will eventually lead, our friend says with a smile that goes back years, 'Well, it's like Zen practice: all you can do is believe you are going in the right direction'.

The Shin home

It is breakfast time in the rural household of a Shin temple family.

The daughter, a junior high school student, is late for school. Her mother reminds her to go to the Buddhist altar to say her prayers. The daughter is reluctant ('Oh Mum, do I have to?') but her mother is adamant. The daughter at last gives in. She goes through the motions wondering how long she has to do this 'morning-worship stuff'. On the bus to school it is crowded; people are standing in the aisles; other people are pushing to get on. She is annoyed.

For many people affiliated with the Buddhist temples in Japan, the Buddhist scriptures have no meaning. They count themselves among the spiritually ungifted and have no aspirations for the monastic life. Even the daughter of a Buddhist temple family in the short scenario above takes little interest in the religious customs of her own sect; she feels put out by having to do the meaningless ritual, added to which she is annoyed by the crowded bus. There seems to be little room for religion in modern life. How, then, has Shin managed to etch its teachings deeply on the life of the people?

In 1988, the city of Kanazawa, as well as many Shin temples in and around the city, quietly celebrated its five-hundredth anniversary. This date marks the founding of an autonomous state by Buddhist farmers in north-eastern Japan in 1488. This state is regarded by some historians as the first democratic state in modern history. It came to an end in 1582 when government troops invaded the city and destroyed the main temple that was its centrepiece, as well as the surrounding township.

The Buddhist faith has managed to survive this bitter ordeal through the establishment of independent faith-study groups. These voluntarily organized groups were often independent of the temples. Their focus was on the faith of each individual member. Membership was open to all, Buddhist and non-Buddhist. It would be expected, for instance, that the case of our recalcitrant daughter would naturally resolve itself over time, through her participation in these faith-study groups. It was through affiliation with such groups that the religious understanding of the people matured into the strongly individualistic version of Shin that exists in the region today.

The Shin followers have always had a strong feeling for their religion. Traditionally belonging to the agricultural class, they were the ones who tilled the soil and lived closely tied to the cycle of the seasons. The Buddha in whom they believed was the source of the dynamic force in the living world around them. It was this shared view of life that created a democratic society, which worked at cross purposes with the hierarchical one the fifteenth-century government sought to impose.

The Shin practice is the recitation of the Buddha's name at morning and evening worship. Faith in the Buddha is a common feature of popular Buddhism everywhere. In modern Shin understanding, however, the Buddha is not a divine being that exists 'out there' but a dynamic force working for our salvation who holds all living beings in mind as the object of contemplation. In this sense, he is the ideal practitioner, dynamically engaged in helping others.

The young Shin priest finishes the morning service before the Buddhist altar in the main hall. His wife reminds him of his schedule for the day. Later, he will be attending a meeting to oppose government plans to establish a new nuclear power plant in the area. Most of the members of his group are Buddhists like himself, but social activism is not part of the Shin agenda. He is just doing his part as a concerned individual. There are no heroes, he says. If you are proud of what you are doing as a Buddhist, there's something wrong. You're not reflecting deeply enough on the real motives behind your actions.

The Shin emphasis on the individual is pronounced. Shin faith, unlike blind faith, forces us critically to consider our actions in life. Shin is also paradoxical in that it argues that if people are evil, then they are all the more the object of the Buddha's infinite compassion. The injunction here is not to 'do evil' but to realize that, whether we intend to or not, all our actions have the potential to produce an evil effect. We may abstain from meat, but we still have to take the lives of plants to survive. We may live a celibate life, but we may still do serious damage to others emotionally. In the environmentally compromised world we live in, this comment on human action is important to bear in mind, not just in living an environmentally protective life but also in our personal relations with others.

CONCLUDING REMARKS

The Buddhist way of life in Japan has the potential to develop into an environmentally protective way of life. Love of nature and respect for religious aspiration, elements basic to the Japanese mind, will make the Japanese people responsive to the views of Buddhist thinkers and scholars who have only recently begun to voice their opinions on the environmental issue. The high literacy level of the Japanese people is also expected to play an important role in establishing an environmentally protective society.

6 IT IS DARK OUTSIDE

Western Buddhism from the Enlightenment to the global crisis

Peter Timmerman

In September 1989, a special issue of the magazine *Scientific American* was published with the title 'Managing the Planet'. This is only one sign of a movement toward planetary management that is gathering force as the official strategy for handling the global environmental crisis. Within the next twenty years, unless we make substantial changes in our ways of doing and being, we will soon find ourselves on a planet managed according to the principles of bureaucratic rationality and economic growth which have been grinding themselves into the face of the earth for the past 150 years. The power of the human race is now great enough for us to do this. But our power is not yet—nor may it ever be—enough to sustain the results permanently. Our understanding of planetary ecology is so poor, our capacity for limiting our excesses is so limited, and our ability to stun nature into submission is so temporary that, sooner or later, earth is likely to shrug us off into the abyss, and go on about its business.

To find our own place and voice in addressing the crisis—as Westerners, environmentalists, and perhaps as Buddhists—at least three elements of our current situation need to be explored. This essay tries to touch on aspects of each of these elements and is therefore divided roughly into three parts.

First of all, we need to consider how Buddhism was translated when it came to the West: what was the West looking for, and what did it find in Buddhism? This meeting of cultures set the stage for certain themes in the emerging Buddhist environmentalism of the late twentieth century. So we begin by looking at the circumstances that awaited Buddhism at the time of its full-scale arrival in Europe and America in the nineteenth century.

Secondly, we need to have some understanding of the environmental movement and the forces—social, economic, political, and psychological—that have helped and are helping, to shape it.

Thirdly, we need to investigate the resources offered by Buddhist thought and practice that can help Western environmentalists find new and effective ways of addressing the crisis.

It is perhaps best to state before going any further that the central theme of this chapter is the disturbing fact that, contrary to the popular view of Buddhism as a 'refuge' from the world, to become a Buddhist today is definitely a political act. More specifically, it is a *geo*political act. If there is a basic premise of our global situation it is that there is no escape from the world. Just as there is no longer an 'away' to throw our waste, there is no longer an 'away' to hide ourselves. The famous picture of the blue ball of earth hanging in black space has become linked with the idea that we must set certain limits on our activities. We are presented with something at once very old and very new: the connection of our daily activities to the sustaining of the vast, intricate and amazing world around us. This connection is known and celebrated by many religious traditions as 'the sacredness of the ordinary'.

1. ESCAPE AND INSCAPE

When Buddhism arrived in the West in the middle of the nineteenth century, Western culture had, in a sense, been preparing for it for some time. One hundred and fifty years earlier, the Western Enlightenment, spearheaded by the physical sciences, had begun its dissection of nature for the purpose of examining it and reassembling it according to human specifications. The ultimate goal was to free humanity from the constraints of nature. It was the birth of scientific rationalism, a world view that believed everything could be measured and explained according to observable physical laws. There were no miracles and no unseen spiritual forces. The fathers of the Enlightenment asserted that if there was a God, he was like 'the good watchmaker'; he had made a mechanically perfect world and then withdrawn to let it tick away of its own accord.

The rationalism and scientific investigations of the Enlightenment led directly to the Industrial Revolution and to the Romantic movement which was a reaction against it. The Romantics rebelled against the increasingly mechanized, industrialized, rational world which they saw as isolating individuals from society, from nature and from their own inner power and creativity. It sought to regain the enchantment of a now dis-enchanted world. But although Romanticism was a reaction against many aspects of the Enlightenment, it had inherited a belief in the human individual as the measure of all things.

This was the Western scenario that Buddhism entered, one that was quite different from the Eastern societies in which it had developed. Not surprisingly it was interpreted and misinterpreted in ways which have had a lasting effect on the Western understanding of Buddhism and on why Westerners turn to Buddhism. Perhaps the easiest way of describing what happened is to use an example.

In 1888, a year before he committed suicide, the painter Vincent Van Gogh, then in Arles in the south of France, painted a portrait of himself as a Japanese Buddhist monk, head shaved, eyes orientalized—as he himself wrote—'a simple worshipper of the eternal

Self Portrait Dedicated to Paul Gauguin. Vincent Van Gogh.

Buddha'. Around his head Van Gogh painted a halo of brushstrokes, reinforcing the powerful image of the artist as monk, faithful to the eternal in art. The painting has a claim to be the first piece of truly Western Buddhist art.

Only six years before he painted this stunning self-portrait, Van Gogh had been an apprentice Christian minister, preaching the Gospel among the poor miners of Belgium. He had failed at this career, offending various people by his waywardness and eccentric lifestyle (he insisted on living in even more extreme poverty than those to whom he preached), and he then turned towards art as a new kind of faith. He saw it at first as a new way of capturing the life and sincerity of the poor, and then as a way of capturing the eternal and infinite in the physical world around him. Like other radical painters of his day, Van Gogh was assisted in his efforts by the discovery of Japanese art (carried to the West, so the legend goes, through colour woodcut prints used as packing paper around shipments) and he became wildly enthusiastic about all things Japanese. In a letter to a friend about how the world is to be comforted after the loss of Christianity, Van Gogh writes:

> If we study Japanese art, we see a man who is undoubtedly wise, philosophic, and intelligent, who spends his time doing what? In studying the distance between earth and the moon? No. In studying Bismarck's policy? No. He studies a blade of grass. But this blade leads him to draw every plant and then the seasons, the wide aspects of the countryside, then animals, then the human figure. Isn't it almost a true religion which these simple Japanese teach us, who live in nature as though they themselves were flowers?

For Van Gogh, the artist had become a kind of priest, devoted to showing in his art the divine shining through the world. Oriental ways of perceiving—including Buddhism—could be called upon as part of the attempt to 'see' with a fresh eye, unblinkered by Western rationalism. When the artist used his clarity of vision he transformed himself into 'a better and higher self'. The idea of the artist as a new kind of priest, uniting art and spirituality, can be found elsewhere in this period among many artists, poets, musicians, and others. That is the main reason it is called the Romantic era. In many ways, it can be seen as a time when the erosion of Western faith in Christianity encouraged people to seek spiritual experience through new channels, including works of art and works of nature.

It is important to examine the attitudes of the Romantics more closely, since by a winding route it leads eventually to environmentalism—and especially that part of environmentalism that is most connected to spirituality.

As the 'dis-enchantment of the world' continued, the Romantics found themselves operating on two fronts. On the one hand they exalted individual genius for tapping into the infinite power of deeper consciousness to overcome the split between the self and the world. On the other hand, they searched for momentary 'visionary gleams' in everyday life and unspoiled nature to reunite them with the world. According to this approach, the world in fact has its own sacred meaning that is waiting to be found by the patient or sensitive artist, rather than being a dead material world that has to have a new meaning imposed upon it by the creative imagination of a powerful genius.

One difficulty was that many Romantics were obviously more attracted to the first approach which exalted their own role as generators of meaning and power in the world. This idea is immensely attractive, but it is also immensely dangerous. For one thing, it reinforces the individualism to which Western society is already prone, and for another, it suggests that the infinitely powerful individual is a good thing—a suggestion that has been taken up by people we revere (like Shelley) and those we abhor (like Hitler). It has also strongly influenced the images we use to form our personal ideal.

When Buddhism arrived in the West in the mid-nineteenth century, it was immediately interpreted according to the individualistic views of Romanticism, which was then at its peak. Buddhism was either mistaken for a kind of Hinduism in which the individual self eventually becomes part of an infinite Self, or it was seen as a world-denying religion in which the religious genius overcomes this dismal world by sheer will-power. For the next century, these two alternatives governed what people in the West thought Buddhism was all about. This made it very difficult to understand the true nature of the challenge posed by Buddhism in which the self, not the world, is the stumbling block.

But as well as looking to the individual as the source of all meaning, there remained the other Romantic approach of uncovering 'visionary gleams' or what the poet Wordsworth called 'spots of time', and the novelist James Joyce would later call 'epiphanies'. These are special moments of life when sudden profound meaningfulness radiates out from what appears to be ordinary experience. All we need is to be sensitive to the world so that we can tune in to them. This mystic materialism, to be found especially in nature, is exemplified by William Blake's famous phrase, 'To see a world in a grain of sand, and a Heaven in a wild flower'.

Through the course of the nineteenth century, writers such as the social critic John Ruskin and the poet Gerard Manley Hopkins developed a more and more intense exploration of what Hopkins called 'inscapes'—the sacredness of natural things as they express themselves just by existing. There had already been a long tradition of seeing nature as the 'second book of God'. But what was new was the celebration of the pure 'thereness' of things, the sheer 'graininess' of rocks, trees, and other objects. If one could see this, one could escape into inscape.

Among the earliest environmentalists or naturalists who followed in the footsteps of the Romantics we find many examples of sudden revelations of the spiritual in nature. Some of these revelations are strikingly similar to certain Buddhist 'enlightenment' experiences. This similarity was not just accidental. For instance, the most famous American environmentalist, Henry David Thoreau, was a student of Indian religion, and was the first translator of part of the Buddhist Lotus Sutra into English. The second most famous early naturalist, John Muir, was a powerful spokesman for wilderness experience as a form of religious experience. In his 1911 book *My First Summer in the Sierra* he writes:

> The snow on the high mountains is melting fast, and the streams are singing bankfull, swaying softly through the level meadows and bogs, quivering with sun spangles, swirling in pot-holes, resting in deep pools, leaping, shouting in wild, exulting energy over rough boulder dams, joyful, beautiful in all their forms. No Sierra landscape that I have ever seen holds anything truly dead or dull, or any trace of what in manufactories is called rubbish or waste; everything is perfectly clean and pure and full of divine lessons . . . When we try to pick out anything by itself, we find it hitched to everything else in the universe.

In one ecstatic moment Muir writes:

> I wish I was so drunk and Sequoical* that I could preach the green brown woods to all the juiceless world, descending from this divine wilderness like a John the Baptist eating Douglass squirrels and wild honey or wild anything, crying. Repent for the Kingdom of Sequois is at hand!

The original writings and experiences of Thoreau and Muir have become something of a Bible for the spiritual side of environmentalism (at least the North American variety). Almost single-handedly they established in the popular imagination the idea of wilderness as a vehicle for personal mystical experience. This has developed into the widespread environmentalist belief that a deep, clear, witnessing of unhumanized nature is crucial to our human self-understanding.

* A sequoia is a giant redwood tree.

2. THE RISE OF ENVIRONMENTALISM

Ensuring that there is wilderness in which to have such experiences has been a main theme throughout the rise of environmentalism. However, right from the start, there has been a fundamental tension in the environmental movement which continues today. One pole has emphasized the appropriate management or 'stewardship' of nature and its resources. The other pole has emphasized the protection of nature from human encroachment. This first pole remains a human-centred environmentalism; the second is more eco-centric, i.e. nature-centred. There has been a natural tendency for those interested in spiritual matters to gravitate towards the second of these two poles, and (at least until recently) to be primarily concerned with mystical identification with nature.

Because environmentalism from 1900 to 1960 was primarily concerned with the pole of management, the spiritual dimension of environmentalism more or less disappeared (or went underground). The history of environmentalism during this period was one in which two activities predominated. The first was the creation of parks and conservation areas to counterbalance rapid urbanization and provide people with recreational space. The second was the gradual creation of ecological science, which began to describe the web-like interdependence of natural communities. There were only occasional hints that there might be something less immediately practical and scientific at stake in this new kind of ecological understanding. For example, the ecologist Aldo Leopold in his *Sand County Almanac* (1949) spoke of the need for a 'land ethic', that is, the extension of our human ethics to include the other species with whom we share the land and who support us.

It was only towards the end of this period (1945–60), after the catastrophe of two world wars and the development of nuclear weapons, that certain philosophers began to be troubled by the implications of our new technological capacities. Two events in the early 1960s brought these concerns together with the first stirrings of an ecological consciousness. These were the discovery of strontium 90 in mothers' milk after atmospheric nuclear testing, and the spread of the pesticide DDT through the environment as revealed in Rachel Carson's *Silent Spring* (1963). Out of these events came a new picture of an earth where things were interconnected with other things by complex ecological cycles; and where something dumped in the water or air in one corner of the world might pop up thousands of miles away in something as pure and innocent as mothers' milk; and where man-made artefacts (atomic bombs and synthetically constructed chemicals like DDT) could affect the life-support system of our planet.

These events and the political events of the later 1960s helped generate the beginnings of the contemporary environmental movement. In the same period, various governments began to legislate against the most blatant, local forms of pollution. The stage was then set for, on the one hand, the appearance of left-wing, almost anarchic, Ecology and Green parties in Europe (and to a lesser extent in North America); and on the other hand, the commitment of mainstream political parties to mild ecological reforms, up to and including 'sustainable development'.

It is again noteworthy that during this period the contribution of religious thought or experience to environmentalism was practically nil. The exception was a flurry of accusations launched at the Christian Church for its teaching of 'man's dominion over nature' which was blamed for the West's exploitation of nature. These accusations were straws in a slowly gathering wind that would eventually require religious communities to re-evaluate their traditions.

As the 1980s wore on, it became more and more obvious that the series of local environmental crises were, in fact, symptoms of an emerging global environmental crisis. This was dramatically signalled by the news of the hole in the ozone layer discovered over the Antarctic, and the dire predictions of global warming that finally broke through into the public consciousness in 1988. It was at last recognized that we were at one of the great turning points of human history, and that decisions made (or avoided) by this generation would shape the ecological future of the earth.

Unfortunately, the response to this recognition has been slow, not just because of the reluctance of governments to move faster. For its part, the environmental movement has found it hard to make the transition from simple protest—saying 'no'—to coherent political action—saying what to do. Part of the difficulty can be traced to the Romantic roots still visible in the movement. The side of Romanticism that glories in nature is obviously no problem for environmentalists; but the other side that promotes infinite individualism is now identified with many of the problems that infect modern life. For example, left-wing politics are full of Romantic images of the heroic freedom of the individual (or the society as one great heroic individual) breaking chains, knocking down barriers, and smashing limits. Meanwhile, at the other end of the political spectrum, conservatives who should be interested in 'conserving' have developed a form of free-market individualism that seems to dissolve all traditions in its path, including long-standing traditions of using nature sustainably. This has made it equally hard for 'conservatives' to feel at home in the environmental movement.

These kinds of confusion have made it impossible to identify environmentalists as simply left or right and made the creation of a strong environmental political movement very complicated. For some environmentalists, however, the confusions and complications are signs that the environmental movement is still not truly grounded. This is one reason for the turn towards spiritual traditions as possible guides in finding that true ground.

3. TOWARDS A BUDDHIST ENVIRONMENTALISM

If there is one essential task of environmentalism today, it is to create a new politics which can respond powerfully and adequately to the problems we face. As we have just seen, however, the environmental movement has been prone to confusion.

It should be kept in mind that the major political parties of our time, of left, right or centre, were all created in the nineteenth century, out of the turmoil of the Industrial Revolution. They were based on the need to manage a growing, newly democratized population. Now, at the end of the twentieth century, the visions and theories upon which these parties were based are quite exhausted. We can see this in the widespread decline in party loyalty among voters, and the lack of ideological coherence in the policies of the parties.

What is needed is an original vision which addresses our current situation fully. It should provide a coherent framework of values and ideas based on a definition of a 'person' which puts him or her in a broader context: how a person interacts with and affects (or ought to interact with and affect) other people, other species and the environment generally. It should open up ways in which we can contribute, and are inspired to contribute, to our local and global community. Current political theories, based as they are on nineteenth-century experience, do not do this for us any more. We no longer live in a world of expanding frontiers, unlimited resources, and technological optimism. As a result, when we listen to our political leaders, there is an unreality about them. Their antiquated ideologies have little bearing on what any fool can see happening around us.

One of the most pressing issues is to reassess our popular model of the ideal person which is a cheap copy of the Romantic ideal. What characterizes modern societies is that they are made up of mini-Van

Goghs, that is, millions of citizens who are struggling for self-fulfilment and the fulfilment of their desires according to the ideas of infinite freedom, rebellion, and creativity. This role model, which destines us for frustration and neurosis, has been tied to an economic theory (also hammered out in the nineteenth century) which suggests that human beings are fundamentally self-interested creatures with an infinite capacity to consume, and that our deepest desires are expressed in the things we buy. It is therefore the role of modern society to monitor and manage the inevitable conflicts among greedy beings, and to use the market system to provide us with enough resources to satisfy as many of our desires as possible. The entire system is what could be called 'Industrial Romanticism', or: 'how shopping became a way of life'.

This immensely powerful vision, based on individualism and supported by the exploitation of natural resources, is the driving force of what we see happening all around us. What makes this dynamic especially dangerous is that it is not only unsustainable, but it is so entrenched in the modern understanding of freedom and self-fulfilment that it is almost irresistible. How can we deny people their right to self-fulfilment? Yet how can we survive on a planet of ten billion points of infinite greed?

It is at this dangerous point that the possibility of turning to Buddhism begins to make sense. If we have completely to rethink our way of life as the first step toward a new political vision, then perhaps we can use Buddhism to help us.

The most important construction of modern culture which Buddhism is well-placed to analyse, assess and perhaps dismantle is the Romanticized individual self fed by a mass of technology designed to reshape the physical world. Until now, the environmental movement has mostly focused on the results rather than the causes of this situation, e.g. the belching smokestack rather than the manipulation of desires that made certain objects seem essential to our personal well-being, which made the smokestack happen in the first place. This misplaced focus is, as already mentioned, largely due to environmentalism's confused allegiance to the political visions of the last century.

By contrast, although Buddhism was associated with Western Romanticism, it leads towards a quite different vision. In common with native traditions and some of the other major world religions, Buddhist understanding is of a world in which personal fulfilment is found in interdependence and not independence, where the self is temporary and non-essential rather than the centre of the universe; and where infinite spiritual development is possible within a physical existence that is understood and *accepted* as finite. While some of these themes echo familiar Western ideas, their implications are different and go completely against the grain of the current political options. In particular, by promoting a different vision of what it is to be a person, Buddhism undercuts the aggression driving today's society. Also, by giving a different—and positive—interpretation to the meaning of a life lived according to the limits and constraints of a 'middle way',* Buddhism obviously presents itself as a serious alternative basis for environmental thought and action.

Early on in this essay I suggested that to be a Buddhist today is a geopolitical act for the obvious reason that every one of our acts now adds to or subtracts from the load of human affairs which burden the earth. It is also a geopolitical act because, given the continuing devotion to consumerism, one of the most radical acts we can perform in our society is to consume less, to sit quietly meditating in a room, or to try and think clearly about who we are trying to be. And finally, being a Buddhist is a geopolitical act because it provides us with a working space within which to stand back from our aggressive culture and consider the alternatives.

This working space, with its ways of carefully considering and meditating on what we do, is part of what can be called 'non-violent thinking'. It is likely to be one of the only strategies that will work against a system which is so aggressive in its pretentions to rationality, and which provokes irrational responses in those who are subjected to the working of its powerful machinery. Going to spiritual traditions for solutions to the global crisis will always appear to some people an irrational response—and given the current resurgence in religious fundamentalism there are often good grounds for this accusation. Indeed, this was one of the great accusations of the Enlightenment thinkers referred to earlier. In 1753, the French philosopher Diderot wrote the following in his book *Thoughts on the Interpretation of Nature*.

> Having strayed into an immense forest during the night, I have only a small light to guide me. I come across a stranger who says to me: 'My friend, blow out your candle in order the better to find your way.' This stranger is the theologian.

Having ignored the stranger, the West took a different route over the last 200 years, and pushed back a lot of darkness that needed

* The path to Enlightenment followed by the Buddha which rejected extremes in favour of pragmatic spiritualism. Extreme practices of self-indulgence or self-denial were rejected, as were extreme beliefs which clung to the material world or denied it completely.

pushing back. Nevertheless, there remains a nagging suspicion that the end result of pushing back the darkness will not be finding our way out of the woods, but completely cutting it down to make room for a bright new tree museum.

Buddhism presents itself as a challenging alternative to Diderot's concerns, and to the aggressive ideas that have for so long shaped our thoughts and our actions. The following ancient Zen koan makes the challenge and the alternative explicit.

> Tokusan asked Ryutan about Zen far into the night. At last Ryutan said, 'The night is late. You had better leave.' Tokusan made his bows, lifted up the door curtain, and went out. He was confronted by darkness. Turning back to Ryutan, he said, 'It is dark outside.' Ryutan thereupon lit a candle and handed it to him. Tokusan was about to take it when Ryutan blew it out. At this, Tokusan was suddenly enlightened.[1]

Note

1 'Case 28', Zenkei Shibayama, *Zen Comments on the Mumonkan* (trans. Sumiko Kudo), Harper & Row, New York, 1974.

SECTION C
MEETING THE GLOBAL CRISIS

7 THE ISLAND OF TEMPLE AND TANK

Sarvodaya: self-help in Sri Lanka

Dr A. T. Ariyaratne and Joanna Macy

> Ancient Ceylon [now Sri Lanka], in the centuries before the colonial powers came, was known as the Land of Plenty and the Isle of Righteousness. Beside the vast network of irrigation canals and reservoirs (or tanks) that made the island the 'Granary of the East', rose great temples and stupas of the Buddhist order. Those sacred edifices were constructed from the earth excavated for the canals and tanks, whose construction and maintenance were supervised by the monks. That history lives today in the minds of those Sri Lankans who speak of the inherent relationship between 'temple and tank', or between religion and development.

Sarvodaya signifies the awakening or liberation of one and all, without exception. *Sabbe satta sukhi hontu*, 'May all beings be well and happy', is the Buddhist wish, in contrast to the Hegelian* concept of the welfare of the majority. In a world where greed, hatred, and ignorance are so well organized, is it possible for this thought of the well-being of all to be effectively practised for the regeneration of the individual and society? The answer to that question lies in the lives of hundreds of thousands of village people in Sri Lanka who have embraced the Sarvodaya way to development.

Since 1958, Sarvodaya** has grown from a small group of young pioneers, working alongside the outcaste poor, to a people's self-help movement that is now active in over 4,000 towns and villages, operating programmes for health, education, agriculture, and local industry.

* Hegel was an early nineteenth-century German philosopher. He believed that the views and needs of the individual could only be allowed a voice in society if they were in accord with the interests of society as a whole. In other words, the good of the many took precedence over the good of the few.

** Further information from: Sarvodaya Central Office, 98 Rawatawatte Road, Moratuwa, Sri Lanka.

The Sarvodaya movement has been able to attain such scope and vitality because it has not tried to apply any ready-made solutions or development schemes from above; instead it has gone to the people to draw forth the strength and intelligence that are innate in them and that are encouraged by their age-old traditions.

Some development experts would argue that in our fast-changing world, preparing briskly for industrialization and modernization, tradition has no meaningful role; but the initiators of Sarvodaya believe that, without the understanding of tradition, no new theory or programme forced on the people, however ingenious it may be, will reap the desired results. No programme will be effective if it tries to separate the economic aspect of life from the cultural and spiritual aspects, as do both the capitalist and socialist models of development. With their sole emphasis on the production of goods, they neglect the full range of human well-being. For his or her well-being, the needs of the whole person must be met, needs that include satisfying work, harmonious relationships, a safe and beautiful environment, and a life of the mind and spirit, as well as food, clothing, and shelter.

To meet these needs Sarvodaya has committed itself to a dynamic non-violent revolution which is not a transfer of political, economic or social power from one party or class to another, but the transfer of all such power to the people. For that purpose, the individual as well as society must change. Each person must awaken to his or her true needs and true strengths if society is to prosper without conflict and injustice. From the wisdom embodied in our religious traditions we can find principles for that kind of personal and collective awakening. That is what Sarvodaya has done—listening to the villagers, who constitute 80 per cent of our country, and creating a challenge for them in terms of the ideals they still revere and in words that make sense to them.

Any understanding of Sarvodaya needs to recognize the resources that our movement has drawn from the spiritual and cultural traditions of our people.

In the Buddha's teachings, as much emphasis is given to community awakening and community organizational factors as to the awakening of the individual. This fact was unfortunately lost from view during the long colonial period when Western powers attempted to weaken the influence of the Sangha (the Order of Monks) and to separate the subjugated people from the inspiration to dignity, power, and freedom which they could find in their tradition.

BUDDHIST TRADITIONS AND NATURE

In Buddhist culture, particularly in Sri Lanka, nature was considered sacred. It had been very close to the Buddha in all the important occasions of his life: he was born in a garden; he attained Buddhahood and passed away from this world under the trees; he preached his first sermon in a deer park and from then on nearly always taught outdoors in the shade of trees. On one occasion, when internal dissent divided the community of monks, the Buddha spent his time peacefully in the forest with the wild animals.

Sri Lankans' respect for life resulted not only in the preservation of animals, birds and other creatures but also plants and trees. The popular ritual of Bodhi Pooja, or paying respect to the Bodhi tree under which the Buddha attained Enlightenment by pouring water onto the foot of the tree, has extended to other species of trees. In times of drought, trees were kept alive by the practical ritual.

In one of the well-known suttas* the Buddha speaks of the 'happiness of living in an appropriate environment' (*Patirupa des vasoca*). The environment, whether village, forest, valley or hill is beautified when the right kind of people live there. There should be a perfect balance between the people who live in a place and the place itself. This is achieved when people live with nature without disturbing the flora and fauna; without breaking and injuring the rhythm of life. The idea is beautifully expressed by the description of a Buddhist saint (*arahant*) who is said to go about in the manner of a bee collecting nectar from flowers but not harming them in any way.

The Buddha often used examples from nature to teach. A mind, flickerinnnng, difficult to guard is compared to 'a fish drawn out of water'. A fish taken away from its watery home jumps here and there. It reaches its end quickly. On the one hand the simile evokes the fickle condition of the mind and on the other it suggests the sad end of the fish snatched away from its environment.

In Buddhist stories, the plant and animal world is treated as part of our own inheritance. The stories tell how animals and plants could once talk and respond to human beings. The healthy rapport between plants, animals and humans, underlined by boundless compassion, was the basis of Buddhist life.

Compassion creates the foundation for a balanced view of the entire world and of the environment in which we live. It is only by exercising loving compassion toward all that a human being can

* For sutta see footnote on p. 19.

perfect him- or herself and become a cherisher and sustainer of life. This teaching is immortalized in the story of young Prince Siddhartha (the Buddha before he renounced his worldly life) saving a swan shot down with an arrow by his cousin. Because Prince Siddhartha saved the bird it was judged as belonging to him.

> If life be aught, the saviour of a life
> Owns more the living thing than he can own
> Who sought to slay—the slayer spoils and wastes,
> The cherisher sustains, give him the bird.[1]

Nothing in nature should be spoiled or wasted for wanton destruction upsets the vital balance of life. Destroying natural resources is physical pollution, but psychological pollution can also afflict any society, affecting both human beings and the environment. Buddhist teachings explain how this occurs. If a king or ruler is overcome by feelings of hatred, excessive desire and ignorance, his ministers and officials are affected and infected by it. This travels down infecting

Young Buddha and the swan.

everyone in the power hierarchy until it reaches the common people. From there it affects the environment in which they live, plants, trees and living creatures, until all are destroyed. Such is the power of psychological or spiritual pollution.

Let us be clear, however, that when we speak of tradition and its role in development, we do not limit our understanding to Buddhism. The example and relevance of Sarvodaya would be very restricted if we thought it had meaning only for Buddhist societies. The Sarvodaya movement, while originally inspired by the Buddhist tradition, is active throughout our multi-ethnic society, working with Hindu, Muslim and Christian communities and involving tens of thousands of Hindu, Muslim and Christian co-workers. Our message of awakening cannot be labelled as the teaching of one particular creed. Through the philosophy of Sarvodaya—based on loving-kindness, compassionate action, altruistic joy and equanimity, as well as on sharing, pleasant speech, constructive work, and equality—people of different faiths and ethnic origins are motivated to carve out a way of life and a path of development founded on these ideals.

Dr A. T. Ariyaratne
Sarvodaya President

HISTORY OF SARVODAYA

A lungi-clad young woman greeting her friends as she returns on foot, with her valise, to a remote village . . . families assembling and weaving palm fronds to thatch a roof for a pre-school . . . toddlers learning songs, getting vaccinated, bringing matchboxes of rice to share . . . mothers preparing food in a community kitchen, starting a sewing class, pooling rupees for a machine . . . a procession of villagers with picks and banners heading out to cut a road through the jungle . . . a monk in orange robes calling on government officials in their file-filled offices, inviting them to join the group, and to supply the cement for culverts . . . police cadets in the city coming to training courses on community awakening . . . prisoners released from jail to work with neighbouring families to clear parks and playgrounds for their children . . . school dropouts organizing masonry workshops in a corner of a temple compound . . . monks and laypeople chanting sacred verses as a new community shop is opened, as a mile of irrigation canal is dredged of weeds, as a hand-built windmill is erected and begins to pump . . . while in the temple's preaching hall villagers gather to hear their children sing ancient songs and to discuss the construction of community latrines.

What can such a multiplicity of scenes and actors have in common? Each is a fragment of the larger whole that is Sarvodaya. What weaves that whole together is a philosophy of development based on indigenous religious tradition, that is, on the Dharma.*

It began in 1958 with a group of sixteen- and seventeen-year-old students from Nalanda College, the prestigious Buddhist high school which had been founded in Colombo at the turn of the century by the American theosophist Col. Henry Olcott, and which was noted for its excellent cricket teams. The students' young science teacher, A. T. Ariyaratne, inspired and helped them to organize a two-week 'holiday work camp' in a remote and destitute outcaste** village. Ariyaratne wanted his students:

> To understand and experience the true state of affairs that prevailed in the rural and poor urban areas . . . [and] to develop a love for their people and utilize the education they received to find ways of building a more just and happier life for them.

From the outset they went to learn what the villagers themselves needed and wanted, living in their huts, sharing their diet, working side by side sinking wells and planting gardens, and talking till late at night in village 'family gatherings'. They called the camp *shramadana*, from *dana* (to give) and *shrama* (labour or human energy). The experience was so rewarding that the shramadana idea caught on and spread. Within a couple of years hundreds of schools joined in the practice of giving labour at weekend village camps, and a national Shramadana Movement was under way.

As the students graduated and took leadership as adults, and as monks and others joined them, the Movement fanned out beyond the school system, expanding its focus from an educational effort to a developmental one, where the villagers themselves took the initiative. The Shramadana Movement began to emerge as a village self-help movement outside the official rural development programme. Meanwhile Ariyaratne went to India to learn from the Gandhian experience and particularly from the Bhoodan–Gramdan campaign led by the walking scholar-saint Vinoba Bhave. Since Gandhian ideals echoed his own, Ariyaratne brought back the term that Gandhi had used for the ideal society based on truth, non-violence, and self-reliance; and he named his plan the Sarvodaya Shramadana Movement.

In 1968 Ariyaratne took the bold step of testing the validity of his

* Word common to the Eastern religions of Hinduism, Jainism and Buddhism which means universal law, religion, teachings, duty.

** Untouchables; those who are considered outside and beneath the four castes of traditional Hindu society.

approach and the dedication of his young colleagues by initiating the Hundred Villages Development Scheme in some of the most impoverished of Sri Lanka's 23,000 villages. Capital resources were virtually non-existent, but with help from Dutch and German donors the programme took hold, elaborating and refining its methods of community awakening. This Hundred Villages Scheme, which seemed wildly ambitious at the outset, spread within ten years to 2,000 villages; in the following three years it had reached over 4,000.

During the 1970s, with help from foreign agencies as well as many Sri Lankan well-wishers, Sarvodaya had established a headquarters and main training centre in Moratuwa, near Colombo, and a dozen regional centres where community organizers and extension workers in health, pre-school education, agriculture, cottage industry, and village technology were also trained. It had organized over a hundred Gramodaya ('village awakening') Centres, each designed to service and co-ordinate programmes in twenty to thirty nearby villages. It was on this basis that by 1980 the Movement entered a new stage, one which Ariyaratne sees as moving beyond purely 'developmental' activities to 'structural change'.

Putting to the test its belief in local self-reliance, the Movement undertook a radical and methodical decentralization of power, giving to local centres decision-making responsibilities in programme and budget. In this deliberate relinquishing of central control, which is a rare phenomenon indeed, hundreds of Village Awakening Councils or *Samhitis* have been legally incorporated. Empowered to develop, conduct, and co-ordinate their own developmental programmes, these councils are constituted to include balanced numbers of children, women, and youths as well as men and village elders. Further structural changes have included (1) granting autonomy to major spin-offs of the Movement, such as the Sarvodaya Research Institute, the Bhikkus (Monks) Programme, and Suwa Setha, the homes for abandoned and handicapped children, (2) creating locally-run community shops to increase competition and to lower prices of goods available to villagers, and (3) establishing a Deshodaya (National Awakening) Council to spur non-partisan discussion of national policies and opportunities.

PROGRAMME OF SARVODAYA

How does Sarvodaya structure its developmental activities? To understand its organization we will proceed, as Sarvodaya organizers do, 'from the bottom up'. But it should also be kept in mind that, given the wide variety in local conditions, the vagaries of human nature, the deficiencies of funds, and the Movement's reluctance to impose any schema 'from above', the following sketch, while descriptive of some villages, represents an idealized picture when applied to the Movement as a whole. The sketch is a composite, many villages in which Sarvodaya works do not include the full range of activities it describes.

The process begins when a village invites a Sarvodaya worker to initiate a programme of activity. Note that since the demand for the Movement's help far exceeds the supply, it rarely enters the scene uninvited. As a first step, this worker checks in with the local monk and other key figures, and asks them to assemble a *paule hamua* or 'family gathering' of local inhabitants, sometimes in the school, but more usually in the temple or preaching hall. Here the initial Sarvodaya 'pitch' or message is given, telling about the awakening that is happening elsewhere in the country and inviting the villagers to begin to take charge of their own lives by discussing frankly together their common needs. To focus their discussion, the organizer challenges the villagers to undertake a *shramadana* or shared labour project in which they take responsibility for identifying, agreeing upon, and meeting a specific need. This could be cutting an access road, cleaning a well, digging latrines, etc. A work camp is then organized and in the process of its organization, which takes a month or two and involves many villagers, local task forces are formed. Either before or during that first shramadana camp, which can last from a day to several weeks but is usually a weekend, embryonic Sarvodaya 'groups' coalesce: the youth group and a mothers' group usually form first, with the children's group and farmers' and elders' groups coming later. Sometimes this process is initiated by a Sarvodaya-trained pre-school teacher, but in any case a shramadana constitutes the usual initial organizing mechanism.

With the organizing of on-going groups through a shramadana, the village now enters a second stage where the groups identify their own priorities and initiate their own programme, such as planting a garden or conducting a house-to-house survey. The Movement supplies ideas, contacts, skills, and even credit and materials through its Gramodaya Centres (next level up) and its regional and national facilities. Young people who have demonstrated particular motivation and effectiveness are chosen by their groups to undertake training at the nearest Sarvodaya Institute. This training may be in community organization or in health and pre-school education or in a given technical skill (agriculture, batik, metal-working, book-keeping, etc., depending on what projects are locally appropriate and feasible). It is more likely to be put into use, when the trainees return with their acquired skills, because they are of the village and have been locally selected by their peers. This process, which, beginning with the shramadana, is open to all, permits the emergence of local leadership that is an alternative to the power customarily exerted by the larger landowners and merchants.

As local efforts take root. the Movement's national and regional network has yet other resources to offer, such as practical skills in organizing local marketing co-operatives and savings schemes, in monitoring the incidence of malnutrition and disease, and in creating locally appropriate rural technology. These resources also include legal aid services, library services, the development of community shops, immunization and nutritional programmes in conjunction with state and international agencies, and Shanti Sena ('peace-keeping army') leagues where volunteers are trained in crowd control, emergency first aid, and conflict resolution. When a village has reached the point where its children's, youth, mothers', farmers', and elders' groups are functional, it is ready to incorporate its own Village Awakening Council, which then serves as an autonomous legal entity designing its own developmental programme.

In guiding this development, the Sarvodaya movement relies on its identification of the Ten Basic Needs. Considered essential to human well-being, these are: water; food; housing; clothing; health care; communication; fuel; education; a clean, safe, beautiful environment; and a spiritual and cultural life. This list serves both to guide village projects, giving equal priority to some factors which appear 'non-economic', and to help Sarvodayans put their other wants into perspective. In the light of these fundamental requirements for a decent and worthy life, all other wants appear as motivated by greed, sloth, or ignorance.

Joanna Macy

Excerpts from:

Joanna Macy with introduction by Dr A. T. Ariyaratne, *Dharma and Development: Religion as Resource in the Sarvodaya Self-Help Movement*, Kumarian Press, Inc., West Hartford, CT, 1985.

Dr A. T. Ariyaratne, section specially written for *Buddhism and Ecology*.

Note

1 Edwin Arnold, *The Light of Asia*, Trubner and Co., London, 1885.

8 IN THE WATER THERE WERE FISH AND THE FIELDS WERE FULL OF RICE

Reawakening the lost harmony of Thailand

Edited talks and interviews with Ajahn Pongsak
Compiled by Kerry Brown

The two monks crossing the road glowed like traffic beacons in their fluorescent orange robes, which was just as well: the noise and fumes could not have been worse if they'd been inside a vacuum cleaner. But for many Thais the billowing diesel flatulence that whites out the tropical sky of Bangkok is a sign of its prosperity.

The city beckons international business to its bosom. And it comes, lured by land prices that, at billions of baht, are cheap at twice the price if you're used to Hong Kong, Singapore or Tokyo prices, but not such a bargain if you're a Thai with the average annual earnings of 8,000 baht (£200).

The land boom is nation-wide. As agri-business* sweeps in, the price of land soars out of reach of the majority of the population who have spent their lives bent over it. With bank loans running at 5 per cent per month (80 per cent per annum), villagers are being priced off the land and forced to clear national forest reserve illegally. Every year, thousands pour into Bangkok to live alongside the rats in the muddle of corrugated iron and plasterboard that squats beneath the concrete showcase. Poor and uneducated, their options are few. For many, trying to support families back in the villages, it is a choice of prostitution or the drug trade.

Almost 80 per cent of the jungle which once blanketed Thailand has disappeared in less than 25 years, stripped away to feed the world market which the Thai economy enthusiastically entered as the gateway to Western affluence. For a while the great hope that was going to catapult Thailand from the Third to the First World in a single bound was tapioca. By the time Europe got tough with

* Large-scale company-owned farms using advanced technology and selling to the world market. They are usually single-crop.

import restrictions on tapioca, there was a lot of tapioca in Thailand, not a lot of forest, the farmers were as poor as ever and their soil considerably poorer. People were beginning to notice: the land was eroding, the air was hotter, the rainy season was shorter and when it did come, the whole place flooded.

The economic development schemes came and went but the trees just went. In 1985, as the environmental losses became economic ones, the government put a temporary ban on logging and announced its new slogan, 'Who Destroys the Forest, Destroys the Nation', and its new policy—they would reforest . . . with eucalyptus. It did not need the environmental impact study, which was never done, for it to be obvious that single-species plantations of trees that were not native to Thailand would unbalance the soil and give nothing to the local wildlife, or in the end, to the local farmers.

But eucalyptus is fast-growing, all the happier for a bit of degraded soil and low rainfall and there was a nice plump paper market in the West. The Forestry Department began clawing back community rights on forest reserves given to villages and handing them out to the private sector as 'reforestation' concessions for 'degraded' land at an annual rent of 120 baht (£3) per hectare. As it turned out the private sector was not so different from the public one. The president of the company with more land on 'reforestation' lease than any other was Senator Kitti Damnercharwanit, economic adviser to the Prime Minister and Democrat Party financier.

Senator Kitti's logging company and others were soon bulldozing primary forests to 'reforest' with eucalyptus. The soil was not degraded—not yet. The local people who used the forest for their livelihood were not consulted. The loggers were often armed with rifles. Complaints to the government were met with a warning that the villagers were themselves illegally encroaching by living in a 'protected national reserve'.

But in 1990, the logging by Senator Kitti's corporation went beyond the bounds of even the most indulgent interpretation of the law and cost the nation another 30,000 rai of ancient forest. The scandal brought down ministers and almost brought down the government, which went into a huddle to review its 'reforestation' policy.

But there are those who do not believe that new policies are the solution. Fifty-eight-year-old Ajahn Pongsak's robes are the colour of the earth, not of traffic beacons. He is, in his own words, 'a small monk' who chose meditation in the forest rather than a career in the Bangkok hierarchy of the Sangha, the Thai equivalent of the Anglican Church.

> Our government officials possess the power and the authority to solve the problem of the destruction of the forests and the problems that are being created by this and they should do this at once. But what we see instead is that our natural resources are being erased right in front of our very eyes. Wherever officials see some benefit they can derive from the forest, they use the power and authority which the law invests in them to shield those who are bringing about the problems for society by destroying the forest.
>
> When the villagers go into the forest to get timber for their own personal needs they know that they are breaking the law but on the other hand, when these same villagers are hired by officials to fell trees and saw them up for timber for sale, then no laws are being broken!
>
> Things have reached such a pitch that if we look at just who is responsible for creating such predicaments for the nation by destroying the forest, we find that there are just too many of them these days for it to be possible to prosecute them all. Far, far too many. If they started to arrest people in this province, they'd have to arrest everybody from the governor down to the simplest villager because everybody is breaking the law. So everybody is equally in the wrong. Who should treat whom for the illness?

Pongsak is the *ajahn* (abbot) of Wat Palad, a temple near Chiang Mai in the north of Thailand where he works with villagers of the Mae Soi Valley to reforest and irrigate their rapidly desertifying land; land that the government had declared national reserve and therefore not to be touched by anybody, neither villager nor logging company nor monk, without their permission; land that was deforested firstly by large corporations under government licence and then by the villagers themselves.

The clear views and clear actions of this 'small' monk have provoked some very big reactions. The opium trade, the logging and associated industries, government officials with an interest in one or other of these, the Ministry of the Interior, and foreign aid agencies have all come snapping at his heels. As an alleged communist, arms depot manager, racist and instrument of the devil, he has been the target of police and army raids, black magic and death threats. As a national and international inspiration to the green movement, a revelation of its next move now that politics, modern science and economics are proving too much a part of the problem to be the solution, he has won awards and been celebrated by the media around the world.

He is an engaged Buddhist.*

* Engaged Buddhists are an informal international movement of Buddhists who believe that a spiritual life includes active participation in society to relieve suffering.

> A problem is a problem and there is only one thing to do with problems of every level, personal or social, and that is to solve them. Buddhism teaches us to solve all problems at their cause. If we cannot yet solve the cause, then we must contain the problem, while we search for a cure.

LOST NATURAL HARMONY, LOST MORAL HARMONY

> . . . The times are dark and *siladhamma* is asleep, so it is now the duty of monks to reawaken and bring back siladhamma. Only in this way can society be saved. Siladhamma does not simply mean 'morality' as commonly supposed. You may obey all the *silas* (commandments) in the book, it still doesn't mean that you have siladhamma, which in truth means 'harmony', the correct balance of nature. From this comes true morality, the natural result of natural harmony.
>
> The balance of nature is achieved and regulated by the functions of the forest. So the survival of the forest is esssential to the survival of siladhamma and our environment. It's all interdependent. When we protect the forest, we protect the world. When we destroy the forest, we destroy that balance, causing drastic changes in global weather and soil condition, causing severe hardships to the people.

For thirty years, the Ajahn wandered and meditated in the, then, deeply jungled north of Thailand, home of barking deer, leopards, bears and gibbons. He was engaged in the practice of siladhamma that would take him towards nirvana—ultimate release from *samsara*, this endless world of suffering.

But while the Ajahn sought release, the suffering of the world escalated around him. In the 1970s, a tobacco company and a timber company were given concessions by the government for selective logging in the Mae Soi Valley. They were supposed to leave the established trees. They didn't. The government issued no penalty.

Seeing the valley, which had provided them and their ancestors with building materials, fuel, medicine and supplementary food, disappearing virtully overnight, the local people also began to saw off the branch they were standing on. They stockpiled wood, made charcoal to sell, hired themselves out as loggers or cattle grazers, or burned forested land as an easy method of hunting the wild animals and encouraging the tasty green shoots that grow in ashes.

The damage was wholesale, but the decisive blow was struck from the top—on the ridges dotted with the huts of the Hmong hilltribes and patchworked with their fields of opium. The Hmong, originally from south-east China, also arrived in the Mae Soi in the 1970s. They had been driven out of Vietnam by the war and out of Burma by the drastic measures taken by the government to stop them clearing the forests on mountaintops, birthplace of the nation's rivers. (In tropical countries, opium only grows over 1,000 metres, but this is where rivers begin as small streams seeping out of the root and humus sponge of the forests.) When the Hmong ignored the warning from the Burmese government to stop destroying the watershed forests, it was followed up by helicopters shooting at their villages.

After that, says Ajahn Pongsak:

> The hilltribes poured into Thailand. Most people still believe that the bulk of opium comes from the Golden Triangle. In fact, the Golden Triangle is more or less finished. In the last decade the primary region that's been providing gold to the powerful has been the North of Thailand.

The hill-tops and ridges of the Mae Soi were freely available to the Hmong in their migration because the local hilltribes maintained the tradition of never farming on the watershed. So the Hmong's slash and burn method of farming sheared across the mountainous spine. By the early 1980s, many of the headwaters had dried and the three main streams had dwindled to a trickle. The heat in the valley had noticeably increased and the rainfall decreased. During the shortened rainy season, with few trees to hold the water, it rushed down the valley taking the topsoil with it in a torrent of coffee-coloured foam that silted up the rivers further down stream. Parts of the valley turned to lifeless dust, others were colonized by a toxin-producing grass. The wild animals had gone. Crop yields were down to a third of their pre-1970s level.

> Thirty years ago, the weather here was a lot cooler. That is heaven—the coolness and freshness of body and mind. Now we live in the furnace of hell, the direct result of our destructive ways.
>
> In the old days when our forests were intact 'in the water there were fish and the fields were full of rice' as we say. We were not rich but we were content. There was material or economic peace, so there was spiritual peace. Now it is just the opposite. Deprivation causes unrest, naturally. If you have nothing to eat, you may resort to crime. The forest is the creator of environmental siladhamma, ensuring a healthy harmony in people's lives both physically and mentally. This is what I mean by the Buddhist interpretation of siladhamma or Natural Balance.

MISUNDERSTANDING THE CAUSE, MISUNDERSTANDING THE SOLUTION

Through the 1980s, in the wake of the destruction, there was a flurry of government and foreign aid agencies setting up projects in the Mae Soi, as they did all over the country. One of the more ambitious ones involved a 52m baht (£1.8m) grant from the World Bank to establish a rain-fed reservoir system.

Fish, turtle and lotus. Mongkol Wongkalasin.

Now what happened was that there was never enough rain to fill the fifty or so small reservoirs that had been made so they were never of any use. Then the plan was to level 3,000 rai of land for cultivation. There was some nutrient in the soil when they began, even areas of sufficient size to be suitable for farming. But all this good soil was bulldozed into the gullies to be used as earth-fill and the levelled ground was gravel and sand of very low quality for crop-growing. Nothing that was planted would sprout strongly of its own accord and all the crops needed large amounts of fertilizer.

In fact, when dealing with a limited area such as this sub-district, one wouldn't have needed 52 million baht to develop. At that time, 10 million baht would have been more than enough. The unrecognized cause of the problem is the destruction of the forests and the cause of this destruction is that the villagers do not have enough land on which to support themselves. This does not mean that land is lacking. Nor does it mean that there is no water. There is about 8,000 rai of land which could have been brought into use for farming. When the rain-fed project started up about fifteen years ago there was enough water, too, because the forest was still virtually undamaged. The streams would have irrigated not just 10,000 but 100,000 rai of land and still there would have been plenty of water left over since in those days when one forded any of the three streams they were almost waist deep. The technocrats could easily have diverted some of this water to the project area instead of being dependent on stored rainfall for watering the crops. They should have been thinking in terms of protecting the forest which was still dense and luxuriant.

So here we see plainly that our technocrats do not analyse problems in sufficient depth to arrive at the underlying causes and the contributory factors of the problems that affect our nation in the countryside. So that most of that 52 million baht has drained away into an area of gravel and sand. It was nearly a complete waste.

This is why the villagers will no longer co-operate with the officials, because the methods used have obviously been the wrong ones. Instead of improving matters they have made things worse by having a harmful effect on the generous open-heartedness of the villagers and thus their siladhamma is damaged; it is declining steadily.

UNDERSTANDING THE CAUSE, IMPLEMENTING THE SOLUTION

When siladhamma is harmed in this way and there is no longer belief and respect in the villagers' hearts, a deep-seated feeling arises: that the countryside and those who live in it have no part in the developments that are taking place. . . . The government and the officials are seen as the owners of the country, the authorities responsible for achieving

> progress and solving problems. Whatever the problem they face, people ask the government or some charitable foundation to come in and solve it for them. This is another point where siladhamma is being harmed. So where are the causes? The cause is that our villagers have no initiative in self-help, in co-operating physically as a group for the sake of the common good or for their own communities.
>
> This is the cause and a factor is—we must accept this—the government financial subsidies. This is a factor influencing the villagers against self-reliance; to act only if there is some immediate benefit in it for them. . . . Nowadays in each village there are still bullock-carts and these days also there are pick-up trucks which could be used in conjunction with voluntary labour to grade and surface the roads. The same can be said of small bridges. If all the villagers got together to build one, they could do it in a week or at most a fortnight as long as the construction materials were provided for them. But the officials arrange the budget to have the work carried out. The villagers are only expected to use what is constructed for them. When the thing falls to bits, then it's time to ask for another budget allocation to rebuild it. . . . Things have got to the stage now where they sit and wait for help in building or carrying out every type of community work although they, in fact, have the time to do such work themselves.

In the early 1980s the Ajahn decided to set an example by personal action. He took up residence at Tu Bou's cave near the remaining trickle of a stream and helped the villagers build a new canal to maximize use of available water. But the destruction did not stop and by 1985 it was time to take stronger action. He went to see the heads of all nine local villages and arranged a district meeting. By now the villagers were looking starvation in the face. They listened.

> I told them that the forest is not only the source of national and natural wealth, but also their first home. Their very own house that they live in, that they cherish so dearly, is in fact their second home. Without their first home, they cannot have their second home. Why do they insist on destroying the forest, wasting and squandering its gifts, why do they kill its children, our brothers and sisters so mindlessly? For rural people, the forest is not only their first home, but also their second parent. The land is able to feed them only so long as there is water, and it is the forest that provides water for the land, and so food for them. It is this second parent, the forest that gives them life after their human parents gave them birth. We recognize gratitude to our parents as a great virtue, so how can we live off the forest so thanklessly? A mind that feels no gratitude to the forest is a coarse mind indeed—without this basic siladhamma, how can a mind attain enlightenment?

The Ajahn also advised the villagers in no uncertain terms that the solution to their problems was their own responsibility. He told them:

> This step, conserving the Mae Soi, we must do ourselves. When we get permission, we can start on it immediately. By doing so, we will be contributing to solving a social problem at a local, provincial and a national level. That's because although this is a very small area of our country, if people work on the problem in this way in every village, every district and every region and every province, then the problems that the government are trying to solve will be scaled down. But first, you villagers will have to unite among yourselves. That is something I cannot do for you.

And they did. Since 1985, the 5,000 villagers, including schoolchildren, have established a tree nursery at the Ajahn's meditation centre, terraced eroded hillsides and planted 170,000 seedlings, including 20,000 for fuel and building needs, dug four stream-fed reservoirs, seven kilometres of canals and ten kilometres of firebreaks, laid ten kilometres of irrigation pipes, constructed nine kilometres of dirt road for access, and eighteen kilometres of fencing around newly planted areas. All decisions were collective, as is the responsibility for maintenance. Each family contributes a day's labour of one of its members on a monthly basis. Most of the work has been done without machinery. Four reservoirs were dug with JCBs lent by the Royal Forestry Department (RFD) while the villagers manually found and laid the rocks to line the reservoir banks. The RFD equipment was then taken away for another project and funds had to be found for hiring JCBs to continue the work.

To date, almost 90 per cent of the funds for essential materials such as pipes and fencing have come from Nunie Svasti, an artist, former botany teacher and family friend of the Ajahn. The money came from sales at Nunie's batik studio in Bangkok. No one else was willing to risk funding on an untried project, particularly not one that was without government sanction. Even the Ajahn's many educated disciples who are now civil servants in Bangkok did not help. They were unable to understand that forest restoration was Dhamma* restoration. For them, the promotion of Dhamma meant building temples and providing for the needs of monks. So Nunie's studio provided the funds until the growing demands on her time as the project documenter and paperworker eventually forced her to sell it. Funds were subsequently raised through the sale of a piece of land

* For Dhamma see footnote on p. 37.

she had inherited. The project is also assisted by Bhikkuni Sudhamma, a Chinese nun and disciple of the Ajahn, moving with unflappable efficiency between the tasks of co-ordinator, accountant, photographer, and office manager.

The RFD has contributed some seedlings but many have been germinated in the nursery from seeds painstakingly collected from surviving trees. One of the most generous suppliers of lowland species is the Sacred Grove. This acre of towering trees and trailing vines stands in a white crusty wasteland on the valley floor. The logging all around was ruthless. It is said that the grove survived because everyone who tried to log it fell seriously ill. Permission to take seeds from the grove is always asked of its tall white guardian spirit, who is frequently sighted.

Under pressure from the Ajahn and the villagers, local officials found lowland for the Hmong. The villagers offered to link the Hmong up to their irrigation system, build them access roads and help them move down. The Hmong agreed. And then they changed their minds after further discussion with officials of the Thai–Norway development programme.

'COMPASSION' WITHOUT WISDOM: SUBSIDIZING OPIUM

For the Hmong there is a distinct advantage in being over 1,000 metres above sea level, the magic line above which opium will grow, all the more so since the government and foreign aid agencies began subsidizing crop programmes to dissuade them from opium growing. The Hmong are now growing subsidized crops of cabbages and potatoes on the ridges of the Mae Soi Valley and spraying them with incorrectly diluted pesticides banned in Europe and paid for by the Thai–Norway development project. Their knowledge about the use of pesticides is minimal but there is enough understanding of the consequences for neither the Hmong nor anyone else in the valley to eat cabbages and potatoes. But although they don't eat them directly, the pesticides are washing into the surviving water supply of the valley, poisoning the river life and the people who rely on it.

Furthermore, subsidies included, more land is needed to earn a baht from cabbages and potatoes than from opium so, funded by Thai–Norway, the tree line on the ridges continues to be razed. And in the end, opium and cabbages and potatoes are not mutually exclusive. Beyond the neat rows of cabbages, their coats of crystallized pesticide glistening in the sun like heat-resistant dew, there is a line of trees. Beyond that are fields of opium poppies blowing gracefully in the wind. They are not so very hard to find and occasionally the police do come across them. The locals shrug. After all, everyone gets his cut.

The destruction of the watershed forest has caused a deep rift between the Hmong and the indigenous Karen hilltribes who farm at lower levels and harvest the forest sustainably. With the drying of the water supply, many can no longer support themselves and are forced to work for the Hmong.

In a neighbouring district the problem of Hmong destruction of primary forests was dealt with by the local officials forcibly relocating the Hmong. There was a national and international outcry at the victimization of an ethnic minority.

Ajahn Pongsak has little time for such outcries: 'We are not chasing the hilltribes out. We want them to move to suitable land. We will make sure that they have everything they need with which to make an honest living. What we are doing benefits them too—it's their watershed and country too.'

He is profoundly cynical of the bleat of the officials at Chiang Mai Town Hall.

> They told me: 'Ajahn, we must have compassion for the hilltribes, because we are Thai, we are Buddhist, we cannot do anything drastic. We must be humanitarian.'. . . So we have laws. We have officials to enforce them. But we do not employ any of these because we are humanitarian. Instead, we use persuasion and understanding—for ten to twenty years we have been persuasive and understanding, so much so that now nearly all our forests have gone, as you can see.
>
> The world sinks millions upon millions to root out opium through programmes such as crop substitution. But the fact is, no one should live on the watershed. That is the heart, the lifeblood of the land, it is common property belonging to the whole nation and must be protected above all. And opium only grows at high altitudes. Moving the hilltribes down, even a little lower, would automatically end the cultivation of opium. The question is, *do we want it to end?*
>
> Officials with both the duty and authority by law refuse to exercise the power and responsibility of the law. Instead, to solve social ills, they plead compassionate Buddhist humanity. The siladhamma they cite is compassion, which indeed is one part of siladhamma—one part in a hundred, one part in a thousand. Compassion is not the whole of siladhamma. To solve problems, you need wisdom coupled with compassion.

He is not despondent: 'With or without official help we will solve this problem.'

COMPASSION WITH WISDOM: TEACHING PEOPLE ABOUT THEMSELVES

In the short term, the need is to curb the problem by helping people to understand that their own survival depends on the forests and watercourses and teaching them the skills and knowledge they need to regenerate the devastated land. But the long-term solution lies with the religion that has been the strength of Thailand for centuries.

> Sanctity and holiness have their basis in purity of spirit. This purity still exists in the great mass of the people, especially in rural areas and may be reawakened for this purpose.
>
> If we fail to bring back true siladhamma, our country will not survive. Country, religion, king—all interdependent, all One. These three institutions are most central to the Thai nation, but we forget. Modern society reveres officials and intellectuals who look upon religion as an anachronistic institution rather than the true nature and basis of social purity, which it is. In the past, religion in Thai society has mostly taken the form of faith. That still exists, but siladhamma based on faith is not enough. The true basis of Buddhism is wisdom—the knowledge and understanding of the true worth of nature according to the Natural Law which is Truth. The Natural Law applies to all life that still must concern itself with the material world, that still must do its duty in and towards society. It is the regular discipline of doing our duty that establishes ethical principles in our daily lives and the general life of society.

The Ajahn's spiritual and ethical approach to nature is producing the hard results needed in a crisis. The villages have a five-year plan to plant almost a million seedlings and considerably expand the irrigation system. Funding to help with this has recently been agreed by the World Wide Fund for Nature. Environmental education for villages, government officials and other monks, a cornerstone of the Ajahn's work, will also expand and be run from a permanent nature centre with the help of a Ford Foundation Grant.

> I am teaching the people about themselves: how to have siladhamma, how to take responsibility for themselves and for society. Knowing their own position in society, having a right understanding of the situation, knowing and performing their duty to their village and society according to Buddhist perception, this is siladhamma. This is balance.

Around the country, there are now scores of monks following Ajahn Pongsak's example in their own communities. In November

1990, at the Dhammanaat Foundation's* fourth environmental seminar for monks, a monks' conservation association was formed. Since then the group has met at ecologically threatened sites, discussed the problems and taken appropriate action. The site of a proposed dam in the south, which will flood the only large tract of true rainforest left in Thailand, now has a settlement of monks which is maintained on a rota basis.

According to Nunie Svasti, 'Any flooding of the area will kill the monks, the worst crime a Buddhist can commit'.

Ajahn Pongsak says with his smile that cracks open his face like a germinating tiger nut:

> Dharma, the Buddhist word for truth and the teachings, is also the word for nature. That is because they are the same. Nature is the manifestation of truth and of the teachings. When we destroy nature we destroy the truth and the teachings. When we protect nature, we protect the truth and the teachings.

Sources

Transcript of Ajahn Pongsak's talk at the District meeting, October 1985. Trans. Nunie Svasti.

Recorded informal talks by Ajahn Pongsak, Dhammanaat Foundation. Trans. Nunie Svasti.

'The Law is not enough', interview with Ajahn Pongsak by Ing K.

Interview with Ajahn Pongsak, 1990, by Kerry Brown. Trans. Bhikkuni Sudhamma.

For further information

Dhammanaat Foundation UK, Spindles, Westabrook Farm Lane, Rew, Ashburton, Devon TQ13 7EJ, UK

* The Dhammanaat Foundation was established by Nunie Svasti to administer and raise funds for the Mae Soi Valley project and the growing work of the Ajahn.

9 LOOK DEEP AND SMILE

The thoughts and experiences of a Vietnamese monk

Talks and writings of Thich Nhat Hanh
Edited by Martine Batchelor

If we consider our world as a Sangha, a spiritual community, we see when we look around us that there is very little real co-operation between nations. Each thinks primarily of its own interests and then acts at the expense of other nations. Sometimes on the international scene we feel that it's not a good Sangha at all. The United Nations frequently seems helpless in resolving the problems that arise in the world community. At the same time we know that we cannot get out, that we belong to this community of nations. We have no alternative but to stay and help make the air more breathable.

In Buddhism, Sangha is a community that practises harmony and awareness. When there is harmony and awareness to some extent in a community, we call it a Sangha. If there is harmony and awareness, the Dharma* is there. And if the Dharma is there, the Buddha is there too. So when there is serenity, understanding, knowledge, a good vision of the Dharma, then we can say that the Buddha is within the Sangha. This is the unity of the Three Jewels (Buddha–Dharma–Sangha).

TWO DHARMA DOORS

The Sangha is a family in which people have to help each other. One of the skills of a bodhisattva** is to create a true Sangha. By your practice you show other people that it is possible to become an element of the Sangha. With this kind of determination you do not opt out of situations; you stay.

* For Dharma see footnote on p. 37.
** For bodhisattva see footnote on p. 6.

To help achieve such a Sangha in the West, two main Dharma doors should be opened now. The first is the Abhidharma door, the Buddhist psychological approach to the problem of internal peace. People need to know the way to transform their internal formations, to relieve their suffering inside. For this we have to know how to work with our own minds. Meditation should be based on the understanding of mind. Enlightenment should be considered as something very practical; for enlightenment is always enlightenment *about* something. And such insight is not just knowledge but true understanding.

Understanding is fluid, flowing like water, while knowledge is like blocks of ice that prevent the flow. Such is the difference between knowledge and understanding. Only by skilful use can knowledge help understanding. If we stick to knowledge alone, we may be caught in it. The essential thing is to understand. In order to grow one has to transcend such knowledge. It is like climbing a ladder. When arriving on the fifth step we might think we are very far from the ground, that we have found some absolute truth. But then there is no possibility of climbing to the sixth step and so on. The way is always to abandon knowledge and be liberated by understanding. Once we are caught by our knowledge there is no future.

The second important Dharma door is the Buddhist approach to harmony and peace. This refers to the Buddhist spirit of reverence for life. Not only do we have to respect the lives of human beings, but we have to respect the lives of animals, the vegetable and mineral realms, and the earth itself.

The monk and the flower. Marcelle Hanselaar.

The destruction of our health by polluted air and water is linked to the destruction of the air and water themselves. The way we farm, the way we deal with our garbage, are all related to each other. Ecology in Buddhism should be deep ecology. And not only deep but universal, because there is also pollution in our consciousness. Television, for instance, can be a form of pollution for our children, sowing seeds of violence and nervousness in the field of our children's minds. While we destroy our environment by toxic wastes, television and other media are polluting the consciousness of our children. So we need to practise an ecology of the mind. Otherwise our inner pollution will spill over into the lives of other beings.

Thus the practice of Buddhism in the West should be a practice as a Sangha. This means that we must practise together, extending our practice into the greater Sangha of society as a whole. We need to write articles and give speeches to show a willingness among spiritual people to assist in political matters. Political leaders need to be helped by spiritual people who are strong and firm on the issues of peace. For we live in a time when people panic easily and start war out of nothing. Therefore, politicians have to be in touch with people with a deep sense of calm, a vision of what the world should be.

The French socialist government has made some efforts to show its goodwill in this area. Brice Lalonde, an ecologist, has been invited to be the Minister of the Environment. Bernard Kouchner, who has been very active in the humanitarian field—with the Vietnamese boat people, for example—has been invited into the government as minister of humanitarian affairs. This kind of attitude is positive. We should ask our political leaders to pay attention to our needs; to create a better environment and stop all forms of pollution.

BREATHING MINDFULLY

The kind of Buddhism I have been writing about and practising is called 'Engaged Buddhism'*. Here we do not practise only in the meditation hall, but outside as well, while washing clothes, cooking meals and so on. I always recommend mindful breathing before action. When something happens it is good to breathe mindfully, but it is even better to be breathing mindfully before the arrival of that event. Because if you have practised breathing before something happens, you will be able to receive it with calm and see the situation

* For Engaged Buddhism see footnote on p. 89.

more clearly. Without such calm and deep understanding your action will probably bring about more harm than good. This is why breathing is so important.

There are many people who act without breathing mindfully and thereby cause much trouble to themselves and society. Especially among young people, we see many groups with plenty of goodwill. They agree completely with each other on the analysis of the situation, but when they come to action they fail because they cannot act in harmony. After some time they start fighting with each other and can no longer serve their cause at all. This is because they lack a spiritual dimension to their actions.

Breathing brings this spiritual dimension to action. You should breathe mindfully before, during and after acting. This is what we tried to do during the war in Vietnam. At times we had to go out and collect numerous dead bodies of war victims and bury them collectively. Otherwise the corpses would start to rot, which would be very dangerous for the population. And there was nobody else apart from the monks and nuns to do this. At other times we had to risk bullets and shrapnel to rescue families trapped and starving behind a combat zone. How could we have done these things without being supported by spiritual practice? Breathing before and during the action was important to us. Breathing allowed us not only to be calm but to be aware of what was going on. For when you breathe, you return to the present moment and see what is happening here and now.

Another time we were in Singapore trying to help the boat people. We practised sitting meditation, walking meditation, silent meals, in the strictest way. We knew that without that kind of spiritual discipline, we would fail in our work. It was a very difficult situation; if the police had found out what we were doing, we would have been deported. Yet on the other hand the lives of many people depended on our skill and mindfulness. We had two big boats called *Leapdal* and *Roland* to pick up boat people in the sea and a boat called *Saigon 200* to transport water, food and medicine. We also helped boat people who were refused entry into Singapore and had also been refused by other nations. Under the cover of the night we would try to bring them into the compound of the French embassy. In this way they would be safely arrested by the police and then given refugee status. Otherwise they would be towed out to sea to perish there.

Finally our work was discovered and I was deported. The police came at two o'clock in the morning. They surrounded my office so we could not escape by the back door, confiscated our passports and gave us 24 hours to leave the country. At that time we were caring for more than 700 people in two boats at sea, but now *Saigon 200* was stopped from sailing and we were unable to bring any more supplies. What could we do in such a situation? We had to breathe mindfully. Otherwise we might have panicked or fought with our captors, done something violent in order to express our anger at the lack of humanity in people. For the police who came for us did not seem like human beings. They were incapable of understanding the suffering of the boat people and what we were trying to do. They were just like machines, yet we had to treat them as human beings. What could we do, knowing that in 24 hours we had to leave while 700 people were adrift at sea without food or water? Could we have just gone back to sleep? Neither the French embassy nor reporters would have answered a phone call at that time of the night. So all of us practised walking meditation until morning. Breathing was so important for us then, even before anything could be done.

> Breathing in, I know I'm breathing in.
> Breathing out, I know
> as the in-breath grows deep,
> the out-breath grows slow.
> Breathing in makes me calm.
> Breathing out brings me ease.
> Wth the in-breath, I smile.
> With the out-breath, I release.
> Breathing in, there is only the present moment.
> Breathing out, it is a wonderful moment.

MINDFULNESS VERSES

Another way to help us dwell in the present moment is to practise reciting *gathas*, or mindfulness verses. When we focus our mind on a gatha, we return to ourselves and become more aware of each action. When the gatha ends, we continue our activity with heightened awareness. When we practise with gathas, the gathas and the rest of our life become one, and we live our entire lives in awareness. This helps us very much, and it helps others as well. We find that we have more peace, calm, and joy, which we can share with others.

> Waking up this morning, I smile.
> Twenty-four new hours are before me.
> I vow to live fully in each moment
> and to look at all beings with eyes of compassion.

If you really know how to live, what better way to start the day than with a smile? Your smile affirms your awareness and determination to live in peace and joy. How many days slip by in forgetfulness?

What are you doing with your life? Look deeply and smile. The source of a true smile is an awakened mind.

How can you remember to smile when you wake up? You might hang a reminder—such as a branch, a leaf, a painting, or some inspiring words—in your windows or from the ceiling above your bed, so that you notice it when you wake up. Once you develop the practice of smiling, you may not need a sign. You will smile as soon as you hear a bird sing or see the sunlight stream through the window. Smiling helps you approach the day with gentleness and understanding.

> Walking on the earth
> is a miracle!
> Each mindful step
> reveals the wondrous Dharmakaya.

This poem can be recited as we get out of bed and our feet touch the floor. It can also be used during walking meditation or any time we stand up and walk.

Walking on the earth is a miracle! We do not have to walk in space or on water to experience a miracle. The real miracle is to be awake in the present moment. Walking on the green earth, we can realize the wonder of being alive.

> Opening the window,
> I look out onto the Dharmakaya.
> How wondrous is life!
> Attentive to each moment,
> My mind is clear like a calm river.

Dharmakaya literally means the 'body' (*kaya*) of the Buddha's teachings (Dharma), the way of understanding and love. In Mahayana* Buddhism, the word has come to mean 'the essence of all that exists'. All phenomena—the song of a bird, the warm rays of the sun, a cup of hot tea—are manifestations of the Dharmakaya. We, too, are of the same nature as these wonders of the universe.

> Water flows from high in the mountains.
> Water runs deep in the earth.
> Miraculously, water comes to us,
> and sustains all life.

Even if we know the source of our water, we still take its appearance for granted. But it is thanks to water that life is possible. Our bodies are more than 70 per cent water. Our food can be grown and raised because of water. Water is a good friend, a bodhisattva, which nour-

* For Mahayana see footnote *** on p. 8.

ishes the many thousands of species on earth. Its benefits are numberless.

Reciting this gatha before turning on the tap or drinking a glass of water enables us to see the stream of fresh water in our own hearts so that we feel completely refreshed. To celebrate the gift of water is to cultivate awareness and help sustain life and the lives of others.

Water flows over these hands.
May I use them skilfully
to preserve the planet.

Our beautiful earth is endangered. We are about to exhaust its resources by polluting its rivers, lakes and oceans, thus destroying the habitats of many species, including or own. We are destroying the forests, the ozone layer, and the air. Because of our ignorance, fears, and hatred of one another, our planet may be destroyed as an environment hospitable to human life.

The earth stores water, and water gives life. Observe your hands as the water runs over them. Do you have enough clear insight to preserve and protect this beautiful planet, our Mother Earth?

This plate of food,
so fragrant and appetizing,
also contains much suffering.

The monk and the butterfly. Marcelle Hanselaar.

This gatha has its roots in a Vietnamese folk song. When we look at our plate, filled with fragrant and appetizing food, we should be aware of the bitter pain of people who suffer from hunger. Every day, 40,000 children die as a result of hunger and malnutrition. Every day! Such a figure shocks us each time we hear it. Looking at our plate, we can 'see' Mother Earth, the farm workers, and the tragedy of hunger and malnutrition.

We who live in North America and Europe are accustomed to eating grains and other foods imported from the Third World such as coffee from Colombia, chocolate from Ghana, or fragrant rice from Thailand. We must be aware that children in these countries, except from rich families, never see such fine products. They eat inferior foods, while the finer products we eat are put aside for exports in order to bring in foreign exchange. There are even some parents who, because they do not have the means to feed their children, resort to selling their children to be servants to families who have enough to eat.

Before a meal, we can join our palms in mindfulness and think about the children who do not have enough to eat. Slowly and mindfully we breathe three times and recite this gatha. Doing so will help us maintain mindfulness. Perhaps one day we will find ways to live more simply in order to have more time and energy to do something to change the system of injustice which exists in the world.

I entrust myself to Earth;
Earth entrusts herself to me.
I entrust myself to Buddha;
Buddha entrusts herself to me.

To plant a seed or a seedling is to entrust it to the earth. The plant takes refuge in the earth. Whether the plant grows well or not, depends on the earth. Many generations of vegetation have grown bright and beautiful under the light of the sun to create a fertile topsoil. This topsoil will continue to nourish generations of vegetation to come. Whether the earth is beautiful, fresh, and green, or withered and dry depends on the plants entrusted to the earth. The plants and the earth rely on each other for life.

SEE THE SUFFERING

In the Tiep Hien Order which we founded in Vietnam during the war, we follow fourteen precepts. These are also helpful in enabling us to remain mindful of our own thoughts and actions as well as the plight of others throughout the day.

The fourth of these precepts reads: 'Do not avoid contact with suffering or close your eyes before suffering. Do not lose awareness of the existence of suffering in the life of the world. Find ways to be with those who are suffering by all means, including personal contact and visits, images, sound. By such means, awaken yourself and others to the reality of suffering in the world.'

Some meditation teachers tell us not to pay attention to the problems of the world like hunger, war, oppression, social injustice, ecological problems, etc. We should only practise. These teachers have not truly understood the meaning of the Mahayana. Of course, we should not neglect practices like counting the breath, meditation, and sutra study, but what is the purpose of doing these things? Meditation's purpose is to be aware of what is going on in ourselves and in the world. What is going on in the world can be seen within ourselves and vice versa. Once we see this clearly, we cannot refuse to take a position and act. When a village is being bombed and children and adults are suffering from wounds and death, can a Buddhist sit still in his unbombed temple? Truly, if he has wisdom and compassion, he will be able to practise Buddhism while helping other people. To practise Buddhism, it is said, is to see into one's own nature and to become a Buddha. If we are unable to see what is going on around us, how can we expect to see into our own nature? Is there not some relationship between the self-nature of a Buddhist and the self-nature of suffering, injustice and war? In fact, to see the true nature of nuclear weapons is to see our own true nature.

The fifth precept says: 'Do not accumulate wealth while millions are hungry. Do not take as the aim of your life fame, profit, wealth or sensual pleasure. Live simply and share time, energy and material resources with those who are in need.'

How can we have time to live the Buddhist ideal if we are constantly pursuing wealth and fame? If we do not live simply, we must work all the time to pay our bills and there will be little or no time for practice. In the context of modern society, simple living also means to remain as free as possible from the destructive momentum of the social and economic machine, to avoid modern diseases such as life stress, depression, high blood pressure and heart disease. We must be determined to oppose the type of modern life filled with pressures and anxieties that so many people now live. The only way out is to consume less. We must discuss this with others who have similar concerns. Once we are able to live simply and happily, we will be better able to help others. We will have more time and energy to share. Sharing is difficult if you are wealthy. Bodhisattvas who practise living a simple life are able to give both their time and energy to others.

And the eleventh precept offers the following encouragement: 'Do not live with a vocation that is harmful to humans and nature. Do not invest in companies that deprive others of their chance to live. Select a vocation which helps realize your ideal of compassion.'

Right livelihood implies practising a profession that harms neither nature nor humans, either physically or morally. Practising mindfulness in our work helps us discover whether our livelihood is right or not. We live in a society where jobs are hard to find and it is difficult to practise right livelihood. Still if it happens that our work entails harming life, we should try our best to secure new employment. We should not allow ourselves to drown in forgetfulness. Our vocation can nourish our understanding and compassion, or it can erode them. Therefore, our work has much to do with our practice of the Way.

Sources

pp. 100–2: talk at Plum Village, France, 13 August 1990.

pp. 102–7: extracts from *Present Moment, Wonderful Moment*, Parallax Press, 1990.

pp. 107–9: extracts from *Interbeing*, Parallax Press, 1987.

For further information

Plum Village, Meynrac, Loubes-Nernac, Duras 47120, France

10 A ZONE OF PEACE

Excerpts from the Nobel Peace Prize Lecture of HH the Dalai Lama

Thinking over what I might say today, I decided to share with you some of my thoughts concerning the common problems all of us face as members of the human family. Because we all share this small planet earth, we have to learn to live in harmony and peace with each other and with nature. That is not just a dream, but a necessity. We are dependent on each other in so many ways that we can no longer live in isolated communities and ignore what is happening outside those communities. We need to help each other when we have difficulties, and we must share the good fortune that we enjoy. I speak to you as just another human being, as a simple monk. If you find what I say useful, then I hope you will try to practise it.

The realization that we are all basically the same human beings, who seek happiness and try to avoid suffering, is very helpful in developing a sense of brotherhood and sisterhood—a warm feeling of love and compassion for others. This, in turn, is essential if we are to survive in this ever-shrinking world we live in. For if we each selfishly pursue only what we believe to be in our own interest, without caring about the needs of others, we not only may end up harming others but also ourselves. This fact has become very clear during the course of this century. We know that to wage a nuclear war today, for example, would be a form of suicide; that to pollute the air or the oceans, in order to achieve some short-term benefit, would be to destroy the very basis for our survival. As individuals and nations are becoming increasingly interdependent we have no other choice than to develop what I call a sense of universal responsibility.

Today, we are truly a global family. What happens in one part of the world may affect us all. This, of course, is not only true of the negative things that happen, but is equally valid for the positive developments. We not only know what happens elsewhere, thanks to the extraordinary modern communications technology, we are also directly affected by events that occur far away. We feel a sense of sadness when children are starving in Eastern Africa. Similarly, we feel a sense of joy when a family is reunited after decades of separation by the Berlin Wall. Our crops and livestock are contaminated and our health and livelihood threatened when a nuclear accident happens miles away in another country. Our own security is enhanced when peace breaks out between warring parties in other continents.

But war or peace; the destruction or the protection of nature; the violation or promotion of human rights and democratic freedoms; poverty or material well-being; the lack of moral and spiritual values or their existence and development; and the breakdown or development of human understanding, are not isolated phenomena that can be analysed and tackled independently of one another. In fact, they are very much interrelated at all levels and need to be approached with that understanding.

Peace, in the sense of the absence of war, is of little value to someone who is dying of hunger or cold. It will not remove the pain of torture inflicted on a prisoner of conscience. It does not comfort those who have lost their loved ones in floods caused by senseless deforestation in a neighbouring country. Peace can only last where human rights are respected, where the people are fed, and where individuals and nations are free. True peace with ourselves and with the world around us can only be achieved through the development of mental peace. The other phenomena mentioned above are similarly interrelated. Thus, for example, we see that a clean environment is not sufficient to ensure human happiness.

Material progress is of course important for human advancement. In Tibet, we paid much too little attention to technological and economic development, and today we realize that this was a mistake. At the same time, material development without spiritual development can also cause serious problems. In some countries too much attention is paid to external things and very little importance is given to inner development. I believe both are important and must be developed side by side so as to achieve a good balance between them. Tibetans are always described by foreign visitors as being a happy, jovial people. This is part of our national character, formed by cultural and religious values that stress the importance of mental peace through the generation of love and kindness to all other living sentient beings, both human and animal. Inner peace is the key: if you have inner peace, the external problems do not affect your deep sense of peace and tranquillity. In that state of mind you can deal with situations with calmness and reason, while keeping your inner happiness. That is very important. Without this inner peace, no matter how comfortable your life is materially, you may still be worried, disturbed or unhappy because of circumstances.

Clearly, it is of great importance, therefore, to understand the interrelationship among these and other phenomena, and to approach and attempt to solve problems in a balanced way that takes these different aspects into consideration. Of course it is not easy. But it is of little benefit to try to solve one problem if doing so creates an equally serious new one. So really we have no alternative: we must develop a sense of universal responsibility not only in the geographic sense, but also in respect to the different issues that confront our planet.

Responsibility does not only lie with the leaders of our countries or with those who have been appointed or elected to do a particular job. It lies with each of us individually. Peace, for example, starts within each one of us. When we have inner peace, we can be at peace with those around us. When our community is in a state of peace, it can share that peace with neighbouring communities, and so on. When we feel love and kindness towards others, it not only makes others feel loved and cared for, but it helps us also to develop inner happiness and peace. And there are ways in which we can consciously work to develop feelings of love and kindness. For some of us, the most effective way to do so is through religious practice. For others it may be non-religious practices. What is important is that we each make a sincere effort to take seriously our responsibility for each other and for the natural environment.

It is my dream that the entire Tibetan plateau should become a free refuge where humanity and nature can live in peace and in harmonious balance. It would be a place where people from all over the world could come to see the true meaning of peace within themselves, away from the tensions and pressures of much of the rest of the world. Tibet could indeed become a creative centre for the promotion and development of peace.

The following are key elements of the proposed Zone of Ahimsa (Non-Violence):

— the entire Tibetan plateau would be demilitarized;
— the manufacture, testing and stockpiling of nuclear weapons and other armaments on the Tibetan plateau would be prohibited;
— the Tibetan plateau would be transformed into the world's largest natural park or biosphere. Strict laws would be

enforced to protect wildlife and plant life; the exploitation of natural resources would be carefully regulated so as not to damage relevant ecosystems; and a policy of sustainable development would be adopted in populated areas;

— the manufacture and use of nuclear power and other technologies which produce hazardous waste would be prohibited;

— national resources and policy would be directed towards the active promotion of peace and environmental protection. Organizations dedicated to the furtherance of peace and to

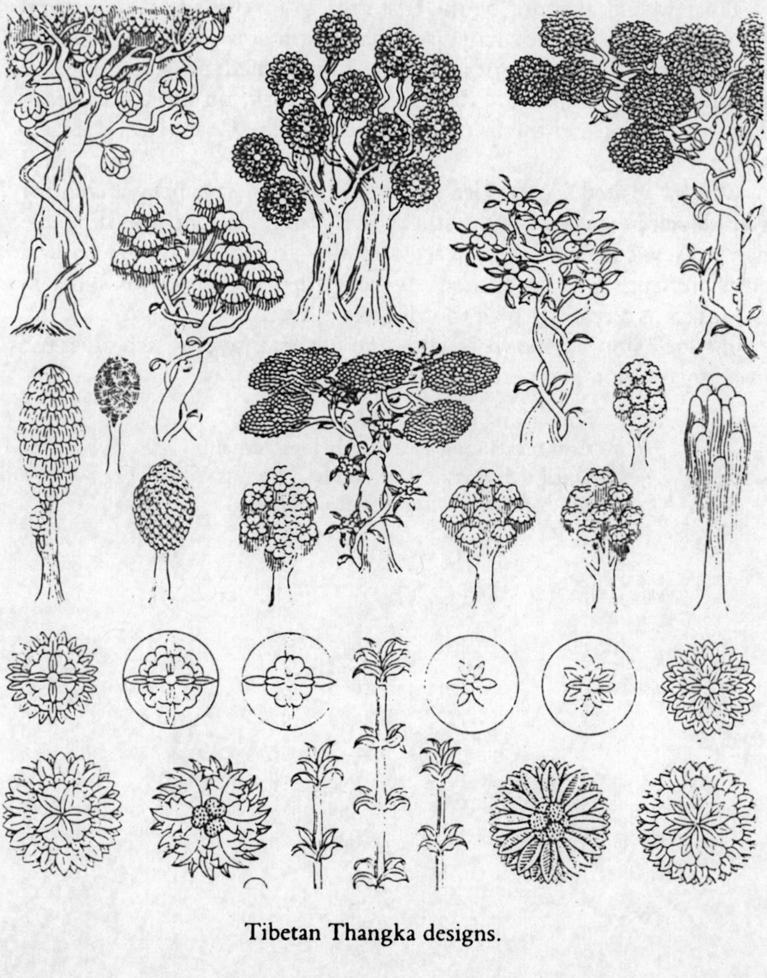

Tibetan Thangka designs.

the protection of all forms of life would find a hospitable home in Tibet;

— the establishment of international and regional organizations for the promotion and protection of human rights would be encouraged in Tibet.

Tibet's height and size (the size of the European Community), as well as its unique history and profound spiritual heritage make it ideally suited to fulfil the role of a sanctuary of peace in the strategic heart of Asia. It would also be in keeping with Tibet's historical role as a peaceful Buddhist nation and buffer region separating the Asian continent's great and often rival powers.

Tibet would also not be the first area to be turned into a natural preserve or biosphere. Many parks have been created throughout the world. Some very strategic areas have been turned into natural 'peace parks'. Two examples are the La Amistad park, on the Costa Rica–Panama border and the Si A Paz project on the Costa Rica–Nicaragua border.

When I visited Costa Rica earlier this year, I saw how a country can develop successfully without an army, to become a stable democracy committed to peace and the protection of the natural environment. This confirmed my belief that my vision of Tibet in the future is a realistic plan, not merely a dream.

In conclusion, let me share with you a short prayer which gives me great inspiration and determination.

> For as long as space endures,
> And for as long as living beings remain,
> Until then may I, too, abide
> To dispel the misery of the world.